Review & Remember Series

Science Fast Facts

Written by Mary Ellen Switzer

Illustrated by Gene Bentdahl

Teaching & Learning Company
1204 Buchanan St., P.O. Box 10
Carthage, IL 62321-0010

Cover design by Gene Bentdahl

ISBN No. 978-1-57310-531-6

Printing No. 987654321

Teaching & Learning Company
1204 Buchanan St., P.O. Box 10
Carthage, IL 62321-0010

This book belongs to

____

Dedication

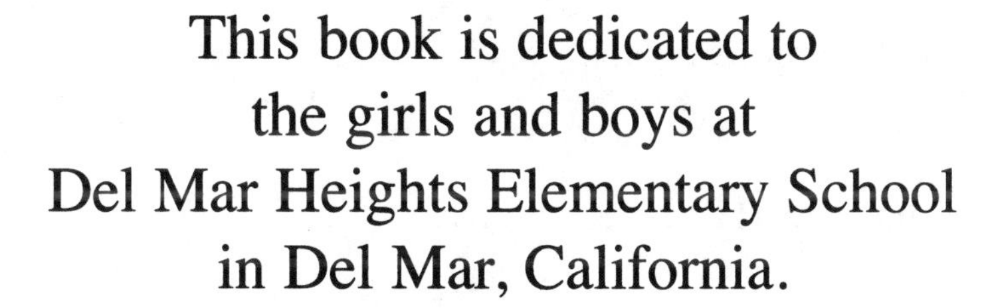

This book is dedicated to
the girls and boys at
Del Mar Heights Elementary School
in Del Mar, California.

Table of Contents

Animals that eat meat are called ____.

True or False? Jaguars are good swimmers.

Herbivores are animals that eat _____.

Tigers live on the continent of _____.

This animal is often called the "king of the beasts."

Zoom! This "sports car" of cats is the fastest land animal in the world.

Lions live in groups called _____.

True or False? Cheetahs need to drink water only about once every three to four days.

What is a female lion called?

Leopards live on what two continents?

Animals

Animals

Animals

Animals

Animals

Animals

Animals

Animals

lioness

Animals

Animals

What do leopards eat?

Are hippos carnivores or herbivores?

True or False? Leopards are not very good climbers.

What African mammal is the world's tallest living animal?

What is the name of the largest of all rhinos?

A giraffe's neck has how many bones?

Rhinos have thick folds of _____ on their bodies.

What does an elephant use its tusks for?

True or False? Rhinos have excellent eyesight.

The loud noise made by elephants is called _____.

herbivores

Animals

birds, deer, gazelles and zebras

Animals

giraffe

Animals

false

Animals 

7

Animals

White rhino. It can weigh between 4000 and 6000 pounds.

Animals

digging, pulling and fighting

Animals

skin

Animals

trumpeting

Animals

false

Animals

An adult male elephant is called a _____ elephant.

How does a bear prepare for its winter sleep?

What animal is the largest of all primates?

Do bears usually live alone?

A troop of gorillas is led by an adult male called a _____.

What is a polar bear's favorite food?

What rain forest animal's name means "man of the woods"?

What color is a polar bear's skin?

It's a world record! Name the smallest bear in the world.

Do polar bears hibernate in the winter?

by eating large amounts of food

Animals

bull

Animals

yes

Animals

gorilla

Animals

seals

Animals

silverback

Animals

black

Animals

orangutan

Animals

no

Animals

sun bear

Animals

Splash! These bears are the best swimmers of all bears.

Animals that have pouches are called _____.

Giant pandas live in the mountains of _____.

The wombat is a slow-moving marsupial that lives in the forests of _____.

These animals live higher up from sea level than any other mammal.

What animal has become a national symbol of Australia?

Name the world's smallest fox.

What type of camel has two humps?

Are foxes nocturnal (active at night)?

Camels lose their fur and grow a new coat in what season?

marsupials

Animals

polar bears

Animals

Australia

Animals

China

Animals

kangaroo

Animals

yaks

Animals

bactrian camel

Animals

fennec fox

Animals

spring

Animals

yes

Animals

What South American animal is the largest rodent in the world?

Microbats use a process called _____ to find food.

This curious rat likes to steal bright objects.

Name the world's smallest mammal.

Name the animal that fools its enemies by pretending to be dead.

Whales, dolphins and porpoises belong to a group of animals called _____.

Bats live in groups called _____.

What is the layer of fat on a whale's body called?

Bats are divided into two main groups. Name them.

Name the loudest of all animals.

echolocation

Animals

capybara

Animals

Kitti's hog-nosed bat

Animals

pack rat

Animals

cetaceans

Animals

opossum

Animals

blubber

Animals

colonies

Animals

blue whale

Animals

megabats and microbats

Animals

What is a young whale or dolphin called?

True or False? A crocodile's teeth stay on the outside of the mouth when closed.

A group of whales is called a _____.

Lizards can be found all over the world except for what areas?

This ocean animal's nickname is the "unicorn whale."

What is the world's largest lizard?

A person who studies reptiles and/or amphibians is called a _____.

A chuckwalla is a large _____ that lives in the desert.

What is the largest reptile in the world?

Do most iguanas live in tropical regions or cold areas?

true

Animals

calf

Animals

North Pole and South Pole

Animals

pod

Animals

Komodo dragon

Animals

narwhal

Animals

lizard

Animals

herpetologist

Animals

tropical regions

Animals

saltwater crocodile

Animals

Most lizards eat _____.

What is the smallest snake in the world?

What is the largest of all sea turtles? It can weigh up to 1000 pounds.

Who spends more time on land—a frog or a toad?

Do turtles lay their eggs on land or in water?

What is the thin skin between the toes of frogs and toads called?

How do tortoises that live in dry places get the water they need?

What is the largest frog in the world?

The heaviest snake in the world is the _____.

Name the biggest toad in the world. It can weigh up to six pounds.

thread snake

Animals

insects

Animals

toad

Animals

leatherback turtle

Animals

webbing

Animals

land

Animals

Goliath frog of West Africa

Animals

by eating cactus and plants that store water

Animals

cane toad of Queensland, Australia

Animals

anaconda

Animals

What is the largest frog in the United States?

Are amphibians warm- or cold-blooded animals?

What frog is named after a jungle cat?

What type of amphibian looks like a worm?

True or False? A toad has longer back legs than a frog.

What does an ornithologist study?

Do toads have smooth or bumpy skin?

Are birds warm-blooded or cold-blooded?

What toad has a "spade" on each back foot that helps it dig?

Changing one set of feathers for another is called _____.

cold-blooded

Animals

bullfrog

Animals

caecilian

Animals

leopard frog

Animals

birds

Animals

false

Animals

warm-blooded

Animals

bumpy

Animals

molting

Animals

eastern spadefoot toad

Animals

Name two birds that cannot fly.

Name the world's largest and most powerful eagle.

What is the largest bird in the world?

What is the national bird of the United States?

True or False? An ostrich is the fastest runner of any bird.

What color feathers does a bald eagle have on its head?

What food does an ostrich eat?

True or False? Eagles hunt only during the day.

What bird lays the biggest egg?

This South American bird is the world's largest parrot.

harpy eagle

Animals

Answers will vary: ostrich, penguin, kiwi and emu

Animals

bald eagle

Animals

African ostrich

Animals

white

Animals

true

Animals

true

Animals

leaves, stems, roots, flowers and seeds

Animals

macaw

Animals

ostrich

Animals

Penguins live only in the _____ Hemisphere.

True or False? The toucan's large bill is very heavy.

Name the largest of all penguins.

What do toucans like to eat?

What do penguins eat?

This record-setting bird is the smallest bird in the world.

Name the national bird of Mexico, which is nicknamed the Mexican eagle.

The humming-bird's long slender bill makes it easy to get _____ from flowers.

What bird is the national symbol of New Zealand?

What bird migrates farther than any other bird?

False. It is hollow and very light.

Animals

Southern

Animals

fruit

Animals

emperor penguin

Animals

The bee hummingbird is about two inches long.

Animals

fish, squid and crustaceans

Animals

nectar

Animals

crested caracara

Animals

Arctic tern

Animals

kiwi

Animals

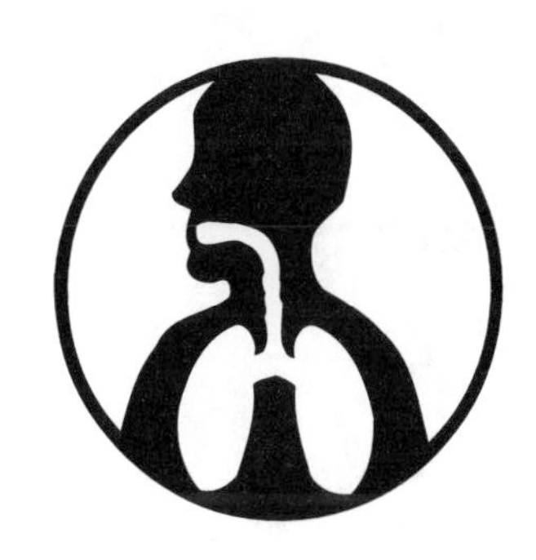

The big three! What three things does your body need to survive?

What do we call the study of the structure of the human body?

Our bodies are made of billions of tiny living units called _____.

What is the body's largest organ?

What human body system is made up of the bones in your body?

An adult's skeleton has how many bones?

This bony part of your head protects your brain.

The human skull has how many separate bones—10 or 22?

What is the only bone that moves in the skull?

A part of your skeleton where two or more bones meet is called a _____.

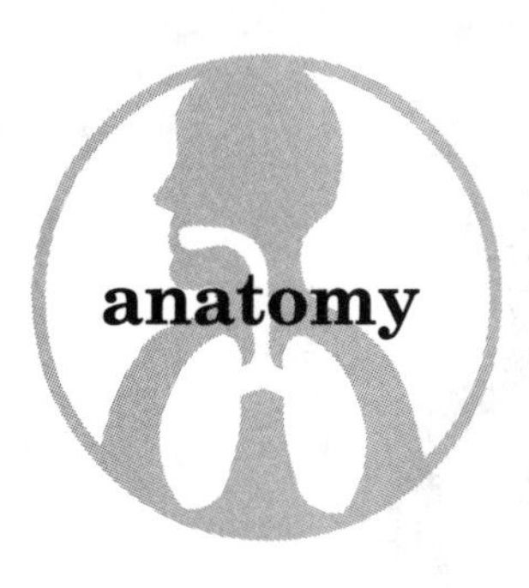

Human Body

Human Body

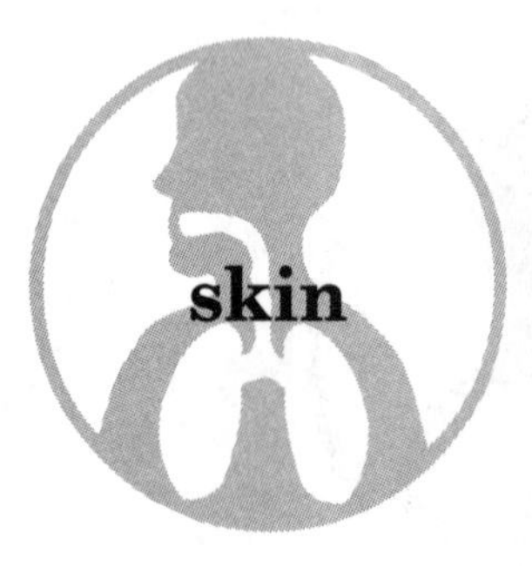

Human Body

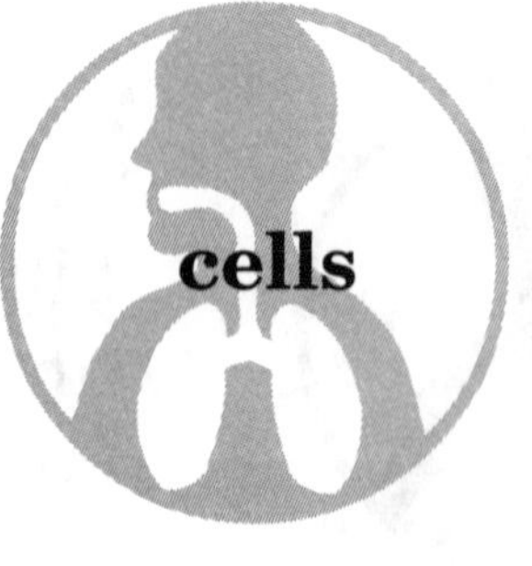

Human Body

Human Body

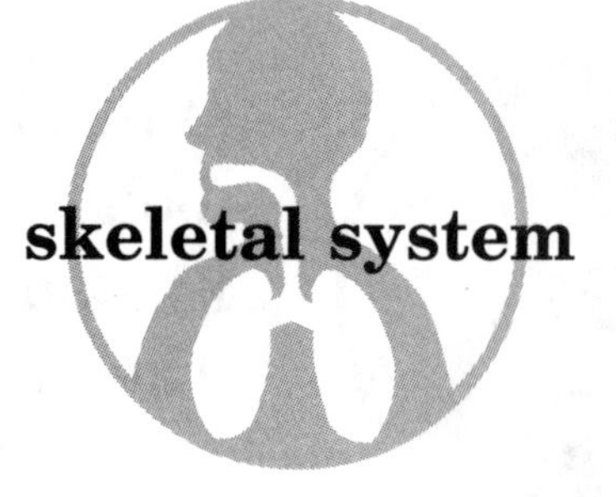

Human Body

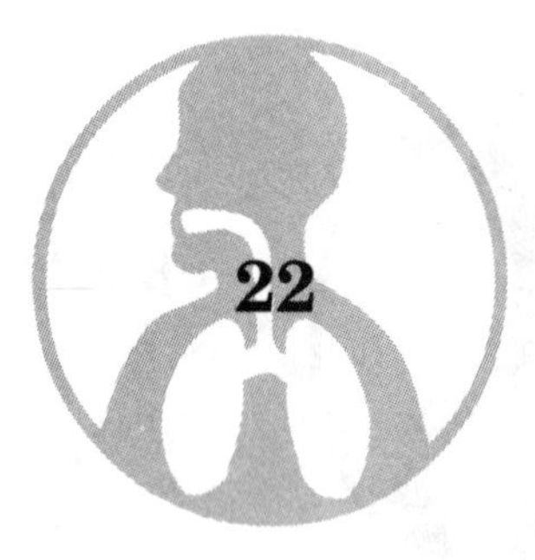

Human Body

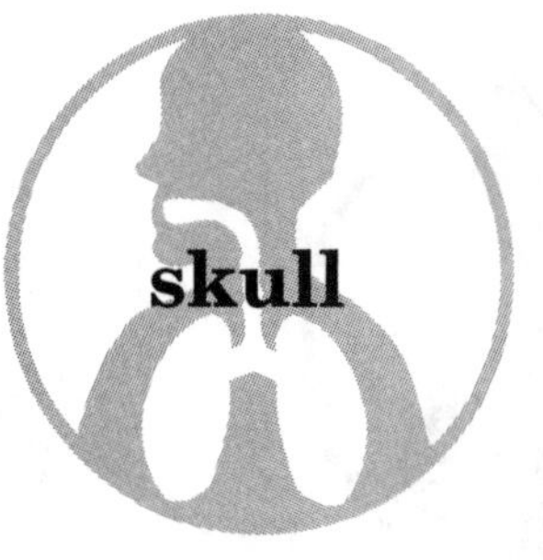

Human Body

Human Body

jawbone

Human Body

Your bones are held at the joints by _____.

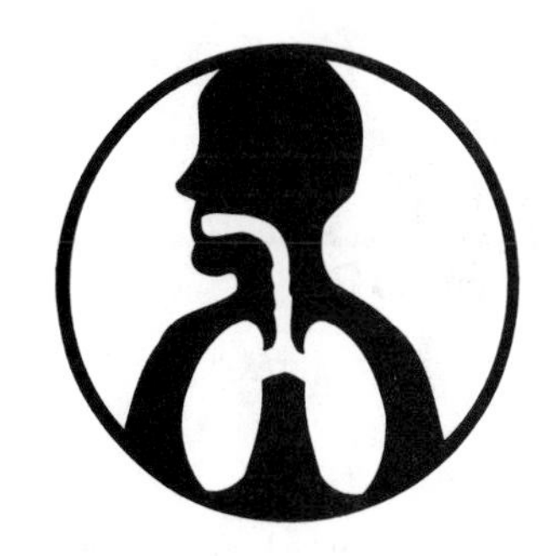

Name the longest bone in your body.

True or False? The tiniest bone in your body is found in your hand.

What is the hardest part of a human bone—the inner or the outer part?

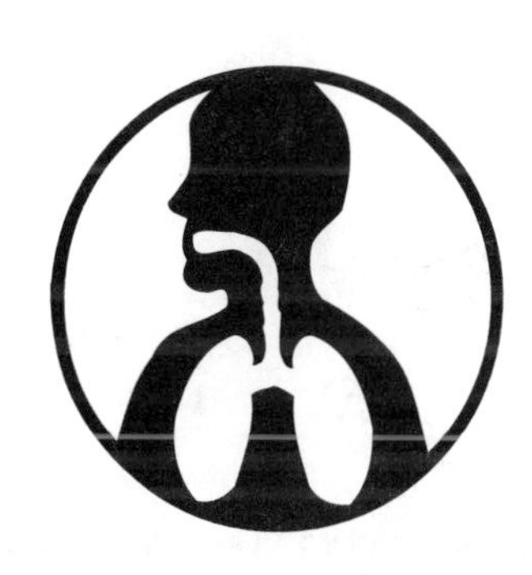

What do we call the soft jelly-like tissue that is found at the center of many bones?

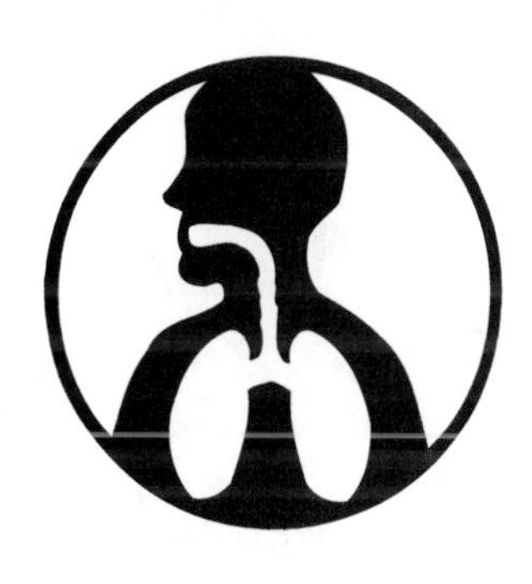

When you are a baby, all bone marrow is what color?

What are the bones that form the backbone, or spine, called?

True or False? Most people have 12 pairs of ribs.

The _____ cage is a collection of bones that protects the lungs and heart.

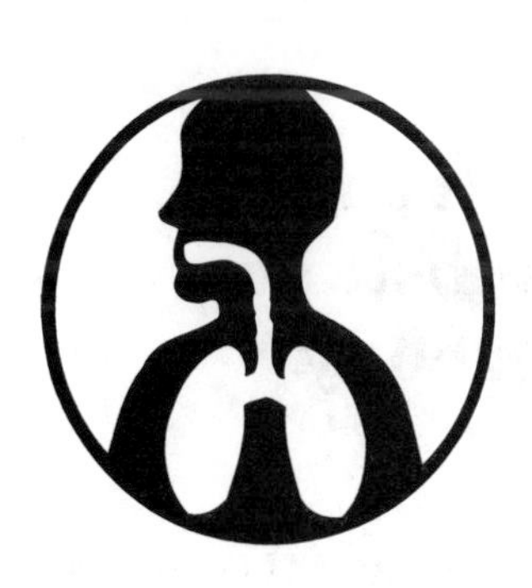

What do you call the hard, white material covering your teeth?

femur or thighbone

Human Body

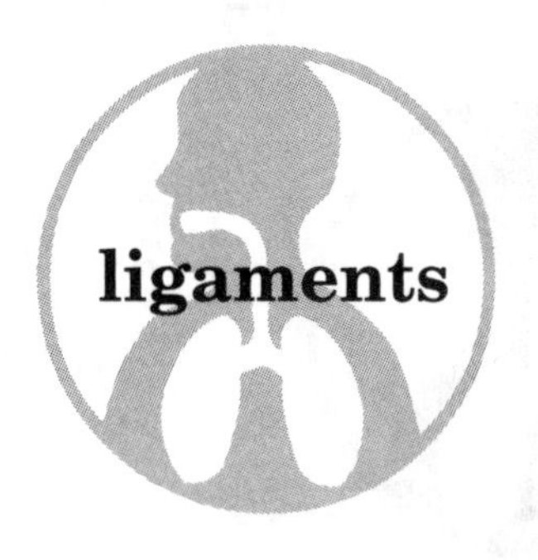

ligaments

Human Body

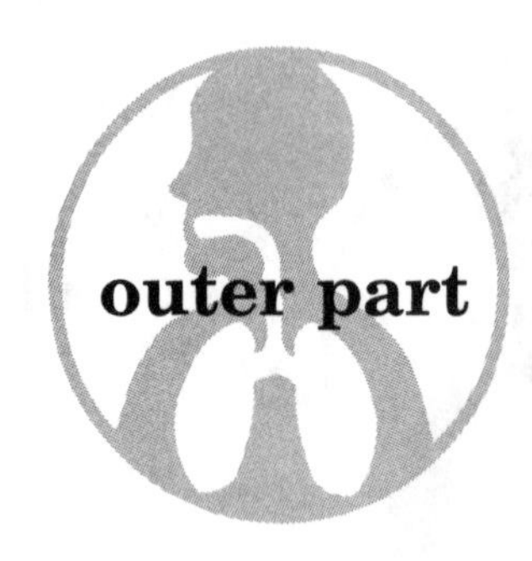

outer part

Human Body

False. The tiniest bone in your body is found in your ear. It is called the stirrup or stapes.

Human Body

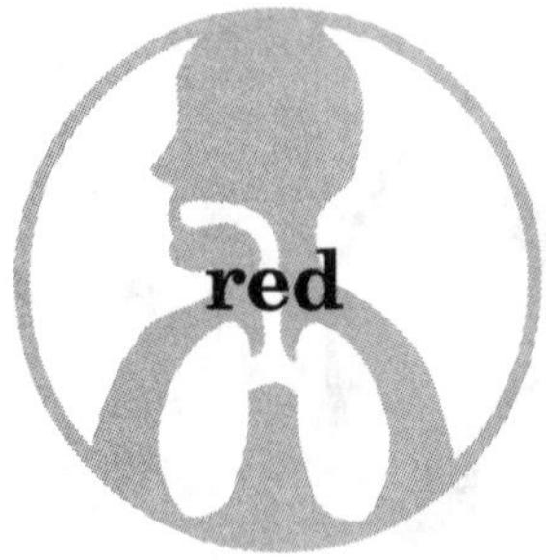

red

Human Body

marrow

Human Body

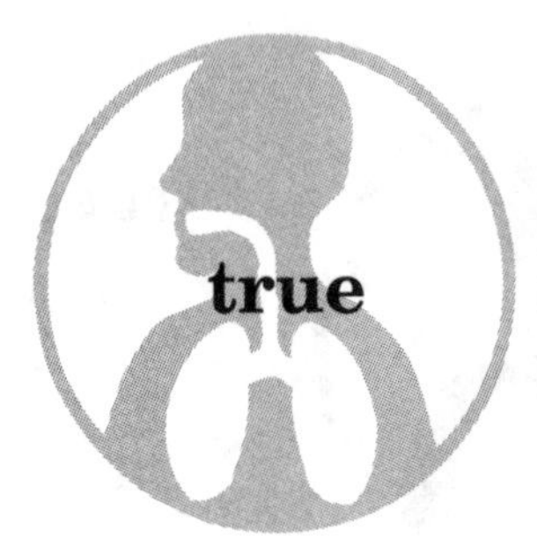

true

Human Body

vertebrae

Human Body

Enamel. It's the hardest substance in your body.

Human Body

rib

Human Body

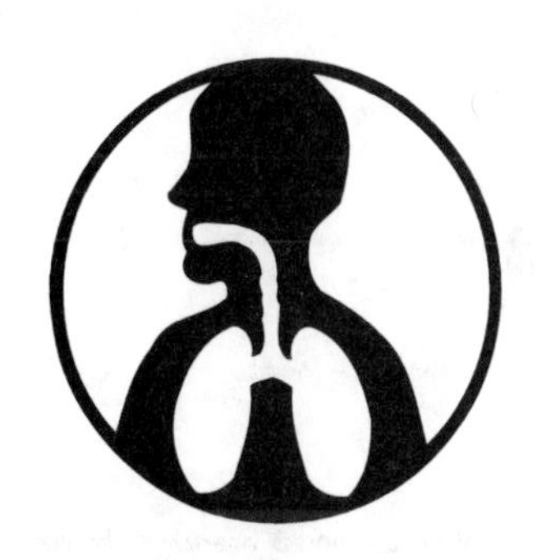

Open wide! The adult body has how many teeth?

The four pairs of teeth in the front of your mouth are called _____.

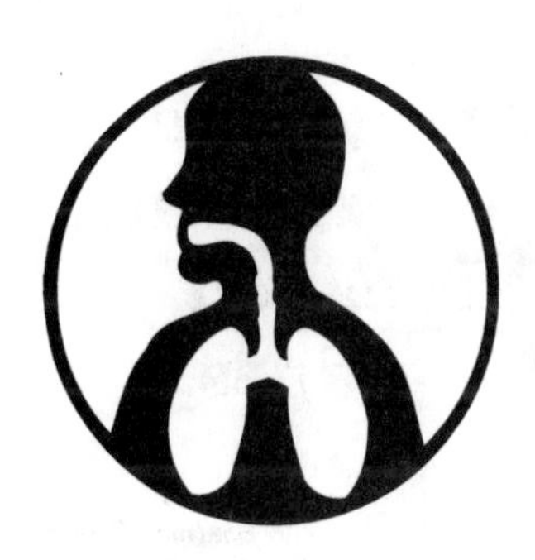

Are molars found in the front or back of your mouth?

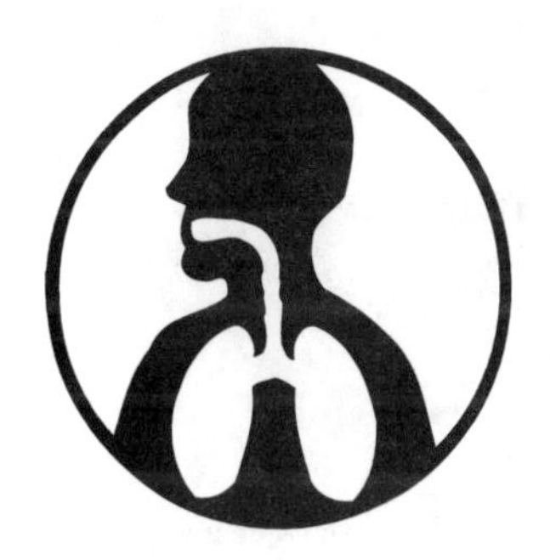

What is a hole in a tooth caused by tooth decay called?

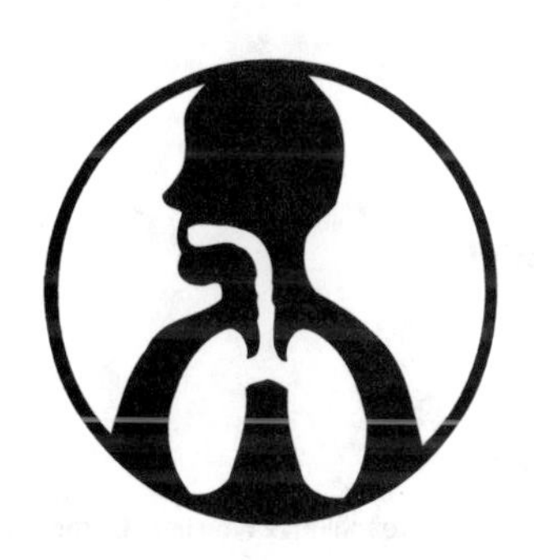

Name the part of the tooth below the gum line.

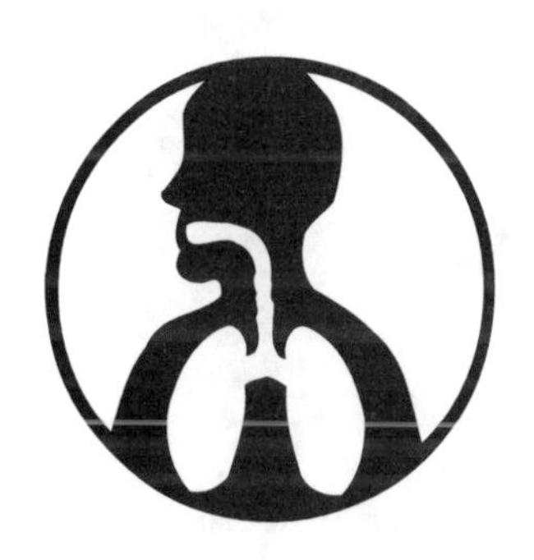

What is the part of your tooth above the gum line called?

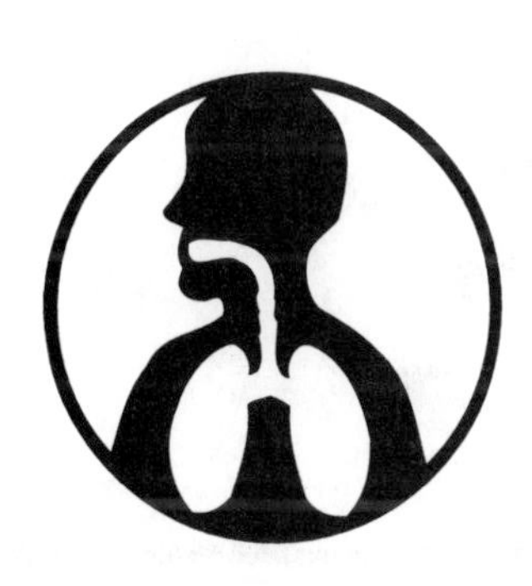

What is the soft pink tissue that wraps around the bottoms of your teeth called?

Your muscles belong to what human body system?

Muscle Power! Muscles are attached to your bones by strong strings called _____.

Are most of your muscles joined to bones?

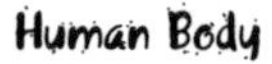

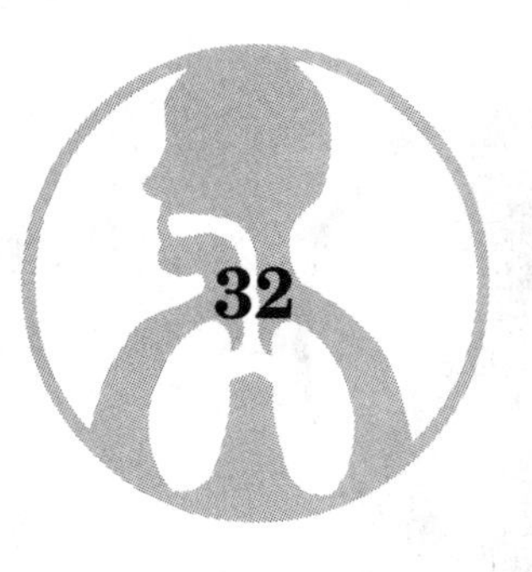

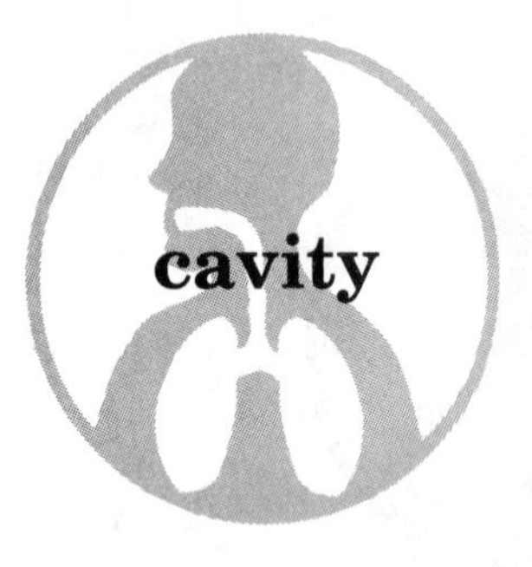

Human Body

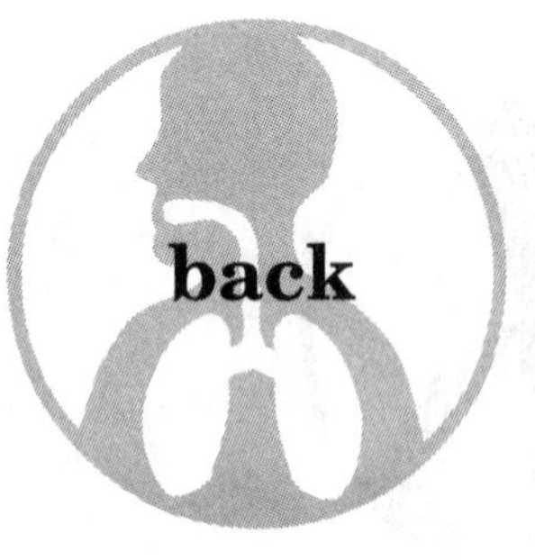

Human Body

Human Body

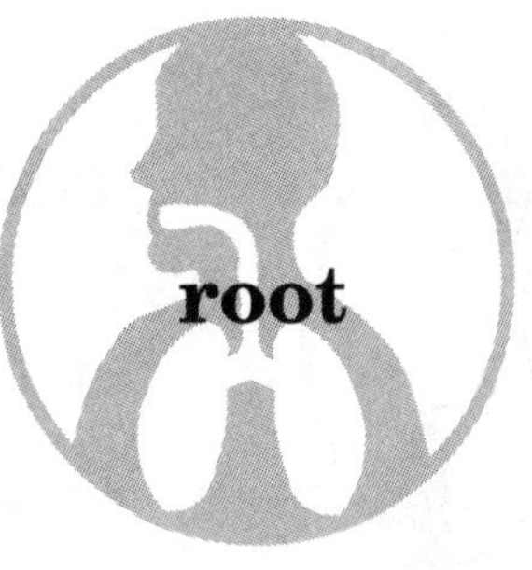

Human Body

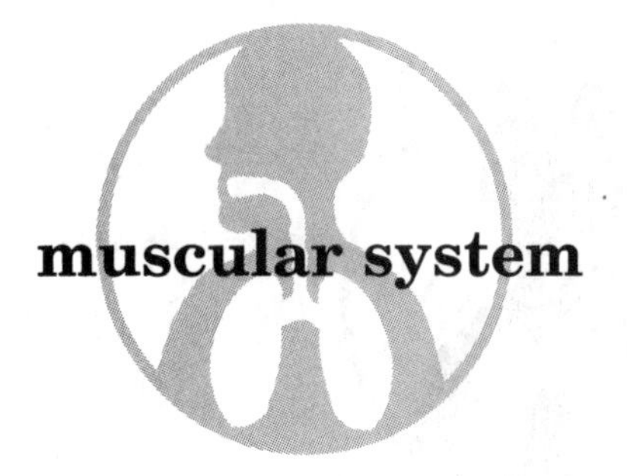

Human Body

Human Body

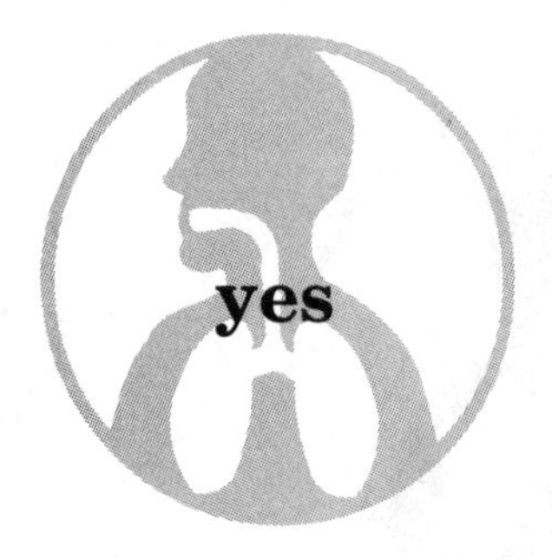

Human Body

tendons

Human Body

These muscles will lift your leg and bend your knee when you walk.

True or False? It takes more muscles to smile than to frown.

What is the name of the big muscle below your lungs that helps you breathe?

The longest muscle in the human body is the _____.

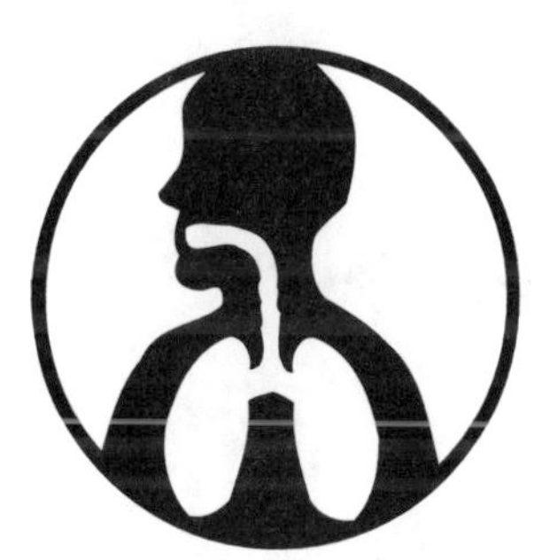

What is the name of the human body system that carries blood all around your body?

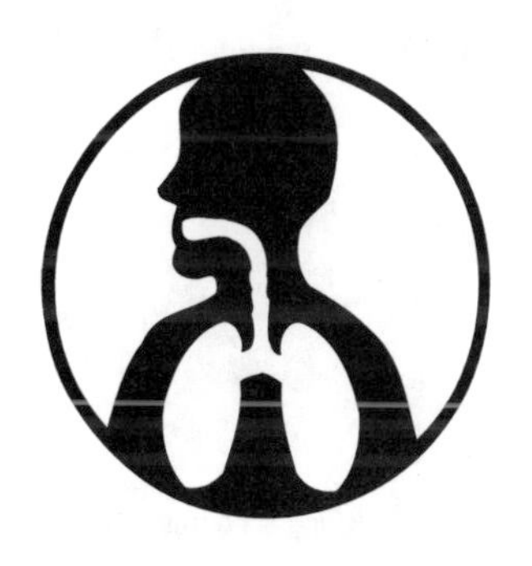

The heart has how many chambers?

Name the two top chambers of the heart.

What are the two bottom chambers of the heart called?

These flaps of muscle keep blood from flowing backward through the heart.

This tube carries blood around your body.

False. It takes more muscles to frown than to smile.

Human Body

quadriceps

Human Body

Sartorius. It runs down your pelvis to your knee.

Human Body

diaphragm

Human Body

4

Human Body

circulatory system

Human Body

left ventricle and right ventricle

Human Body

left atrium and right atrium

Human Body

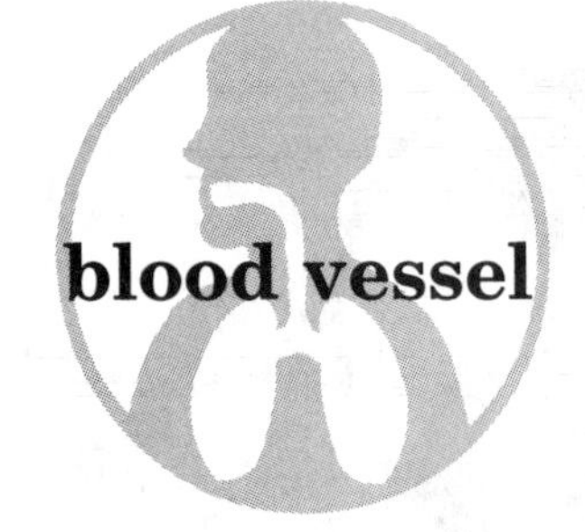

blood vessel

Human Body

valves

Human Body

What do you call blood vessels that take blood away from your heart?

Name the biggest artery in the human body.

Blood vessels that take blood back to your heart are called _____.

Do arteries have thicker, stronger walls than veins?

What are the smallest of all your blood vessels called?

True or False? There are about five million capillaries in your body.

Your blood has two main kinds of cells. Name them.

These blood cells help your body fight diseases by killing germs.

Your red blood cells carry _____ around your body.

What is the name of the system that is responsible for breathing in and breathing out?

aorta

Human Body

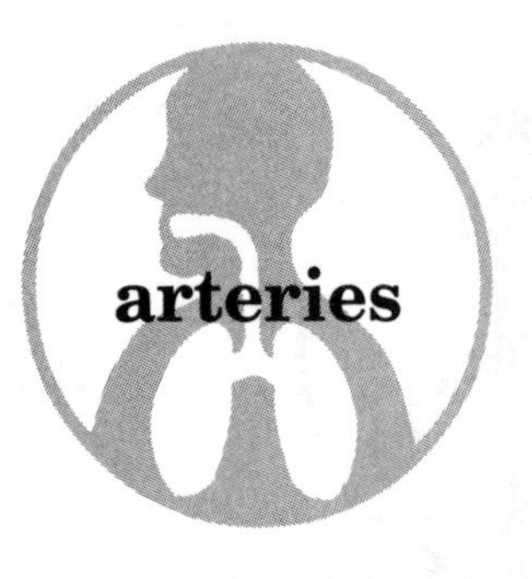

arteries

Human Body

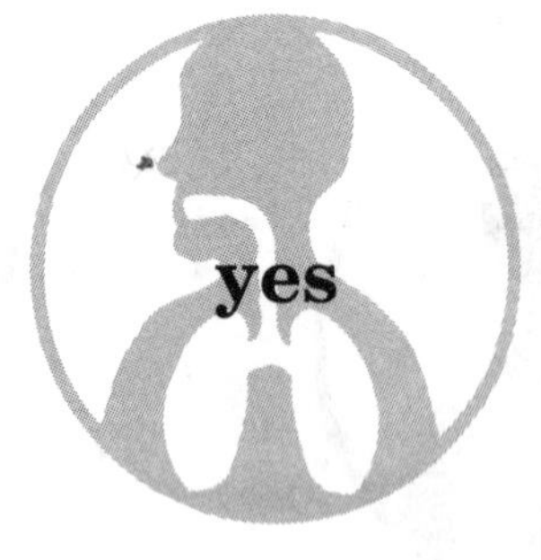

yes

Human Body

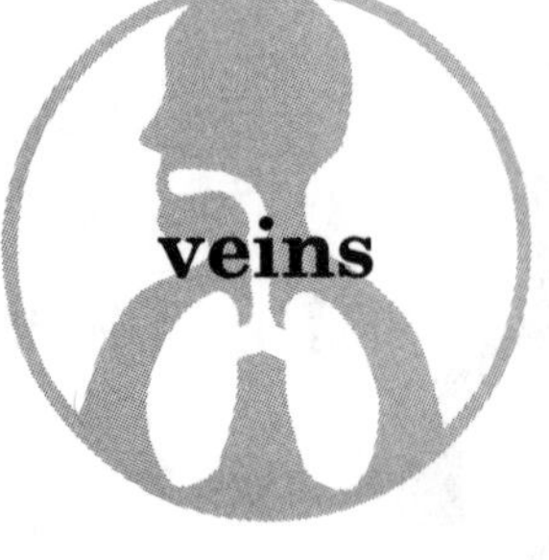

veins

Human Body

False. There are 10 billion capillaries in your body.

Human Body

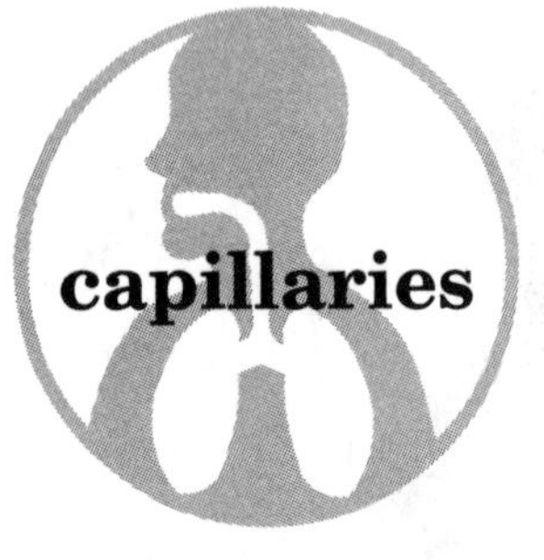

capillaries

Human Body

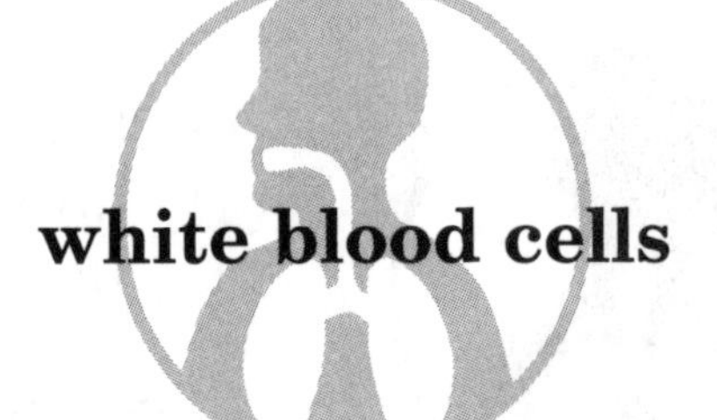

white blood cells

Human Body

red cells and white cells

Human Body

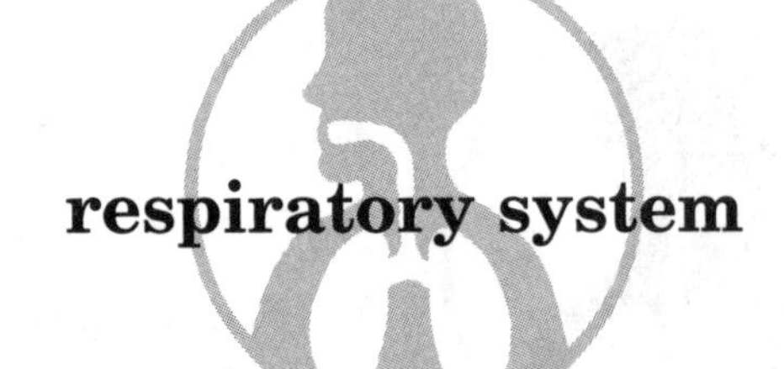

respiratory system

Human Body

oxygen

Human Body

What are the organs in your chest that help you breathe called?

This special "pipe" leads air from your mouth to your lungs and out again.

Air passes through these tiny tubes inside your lungs.

Do children breathe faster than adults?

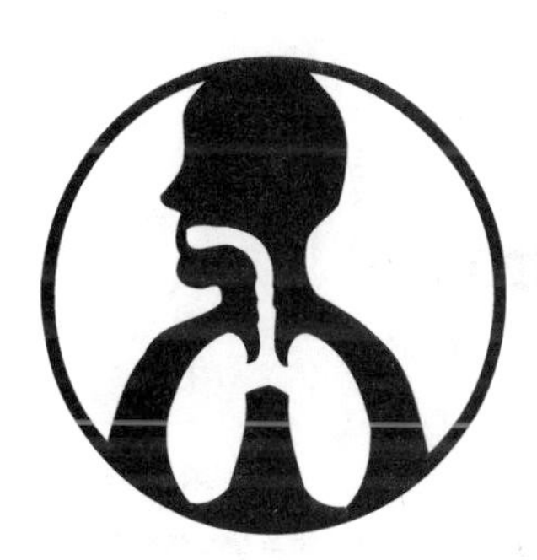

When we breathe in (inhale), we take _____ into our bodies.

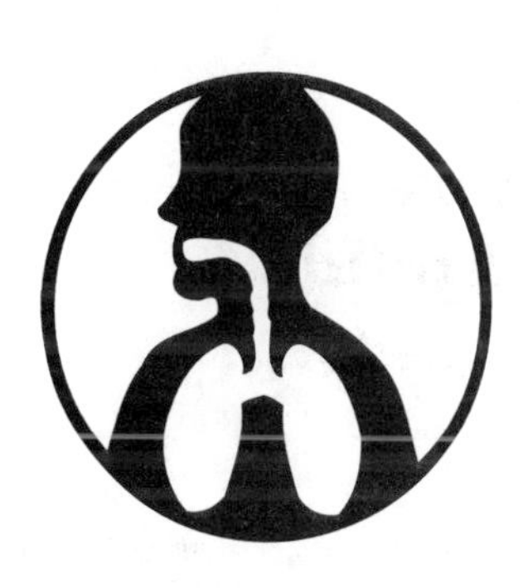

When we exhale, we breathe out _____.

This special tool is used by a doctor to listen to your lungs.

Name the human body system that includes the brain, spinal cord and nerves in all parts of the body.

The human brain is made up of nerve cells called _____.

True or False? Your brain does not work when you sleep.

windpipe or trachea

Human Body

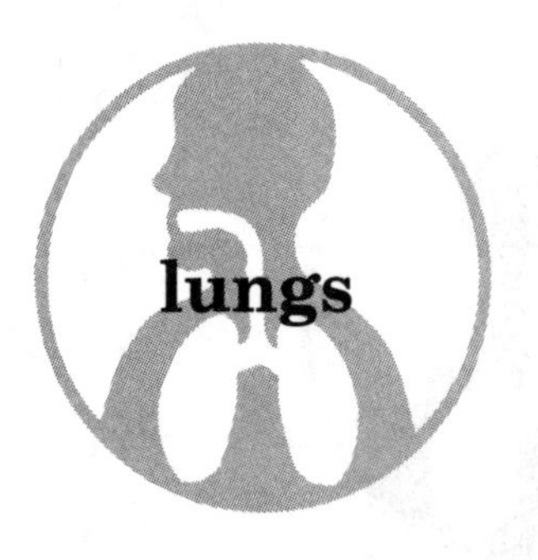

lungs

Human Body

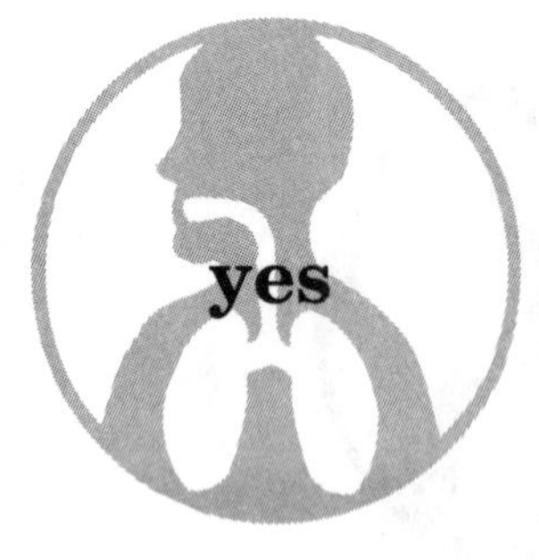

yes

Human Body

bronchial tubes

Human Body

carbon dioxide

Human Body

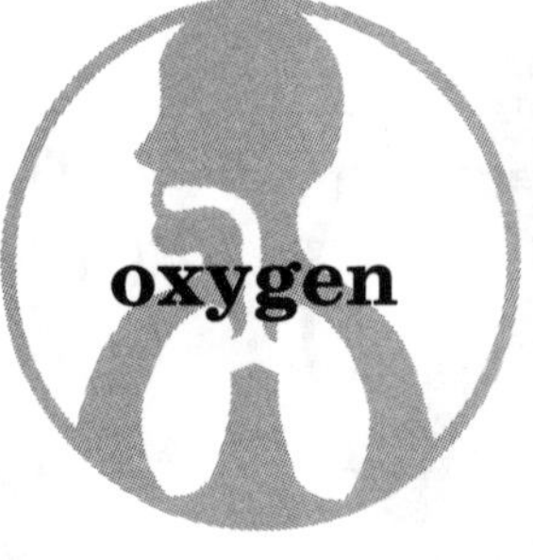

oxygen

Human Body

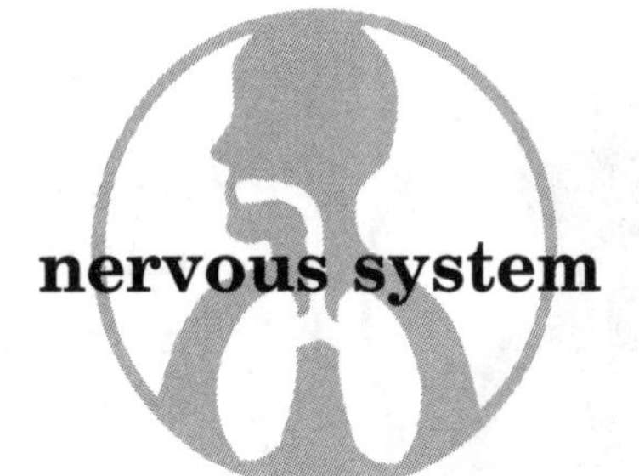

nervous system

Human Body

stethoscope

Human Body

False. Your brain works even when you sleep!

Human Body

neurons

Human Body

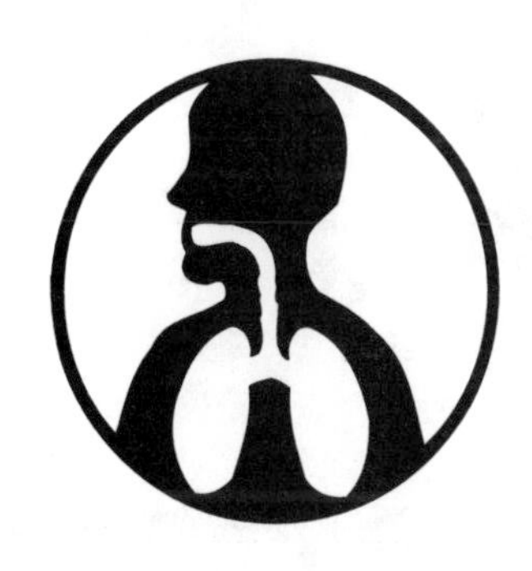

The brain stem joins your brain to your spinal _____.

What is the name of the part of your brain that helps you think?

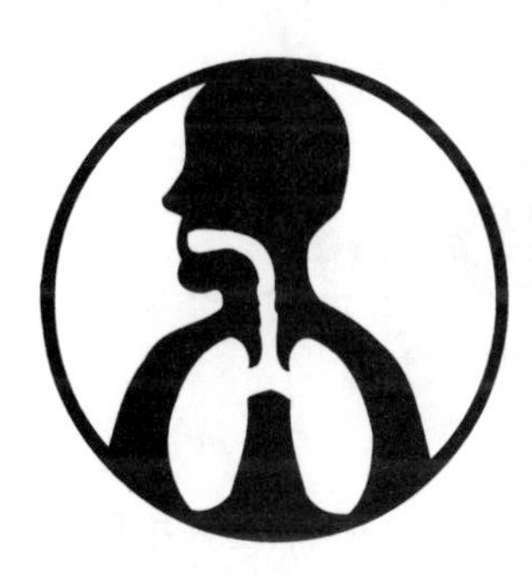

True or False? The cerebellum is the part of your brain that controls balance and coordination.

Every minute nerves send millions of messages to your _____.

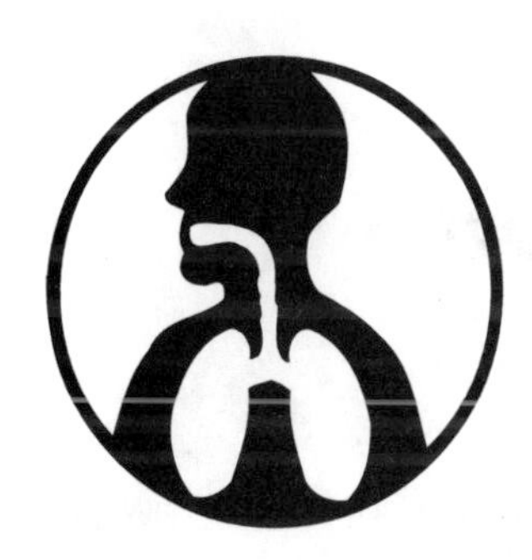

The left half of your brain controls what side of your body?

The right half of your brain controls what side of your body?

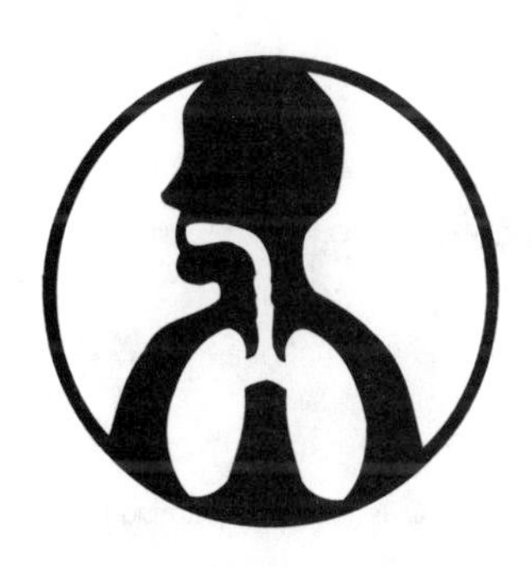

The digestive system moves _____ through the esophagus, stomach and intestines.

The stomach receives chewed food from the _____.

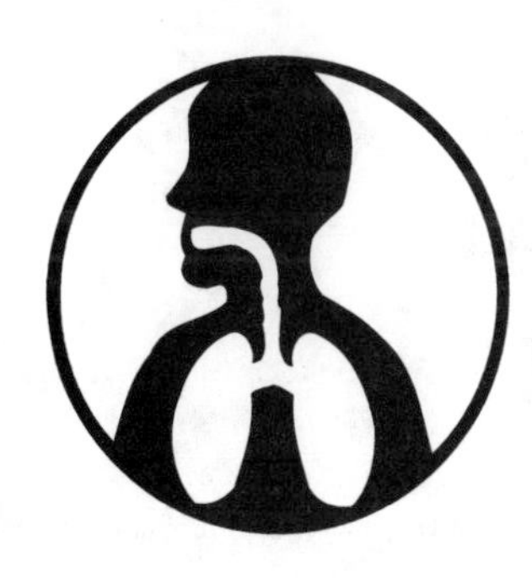

This liquid released into your mouth dissolves food, making it easier to swallow.

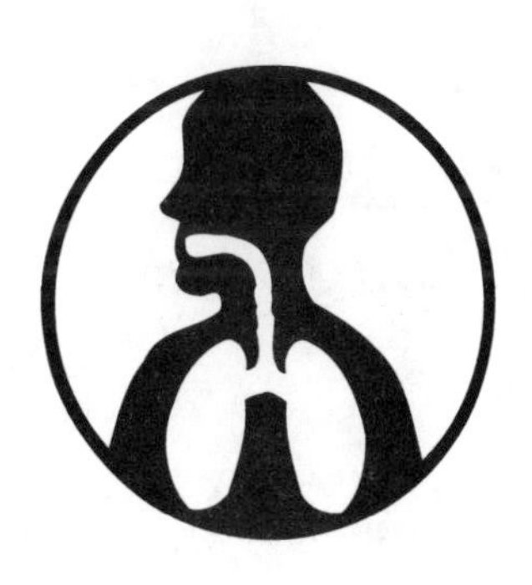

True or False? An adult's large intestine is about one foot long.

Human Body

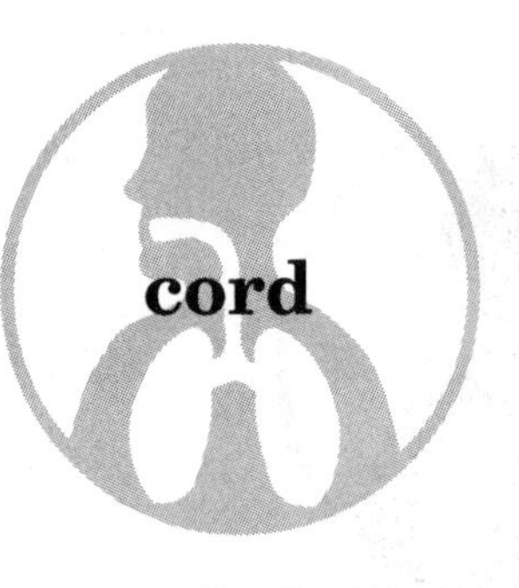

Human Body

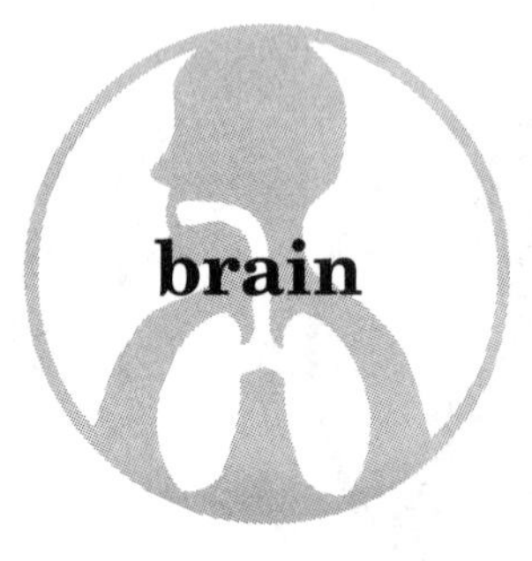

Human Body

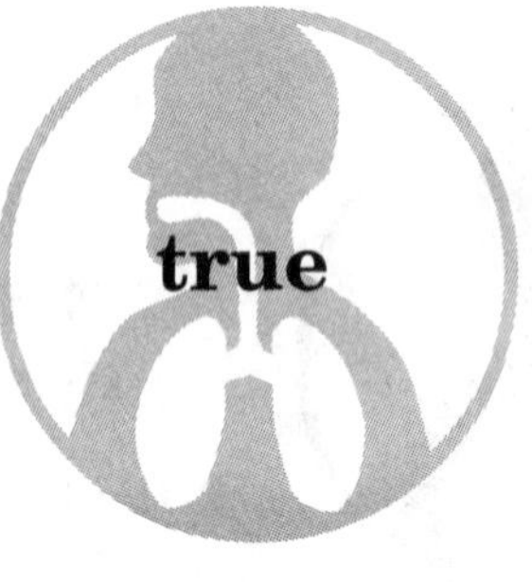

Human Body

Human Body

Human Body

Human Body

Human Body

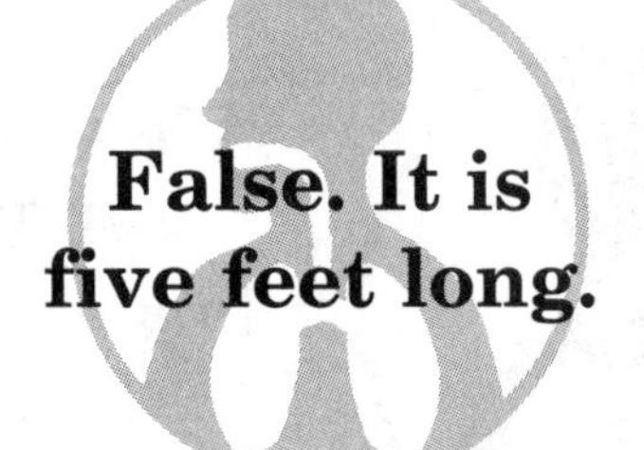

Human Body

saliva

Human Body

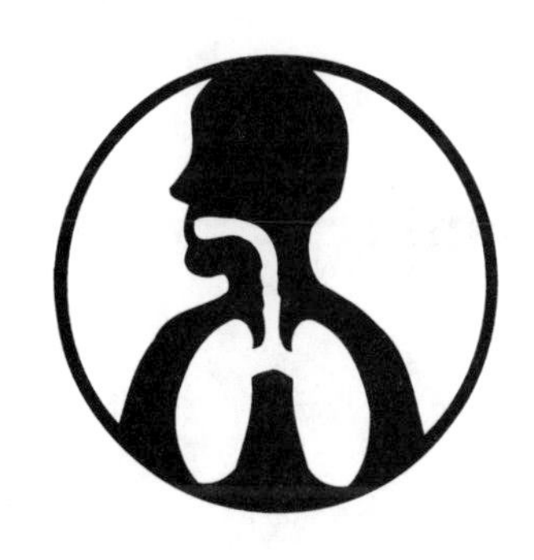

What is the top layer of skin called?

What are the tiny holes in your skin called?

The layer of skin beneath the epidermis is called _____.

What is the layer of skin that covers your skull called?

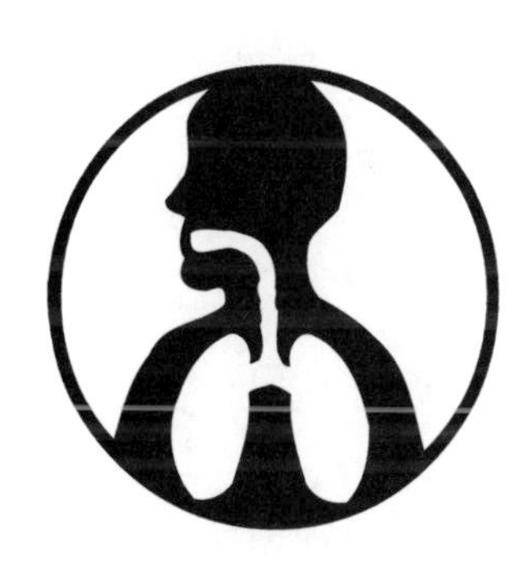

The salty liquid that is made by glands in your skin when you're hot is called _____.

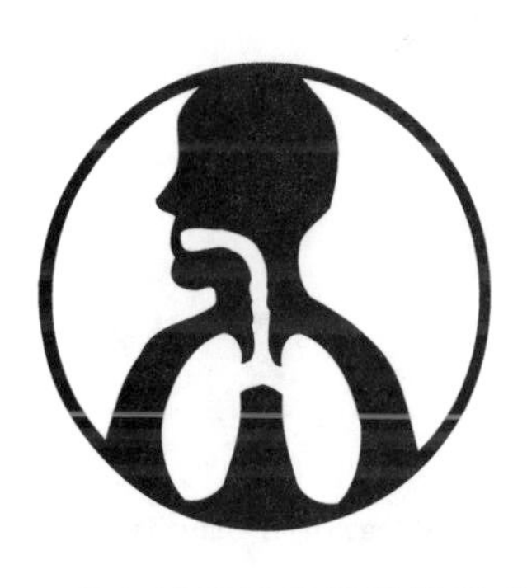

Your _____ protect the ends of your fingers and toes.

True or False? Your hair grows about a half of an inch each month.

Each hair grows out of a tiny tube called a _____.

The Big 5 Challenge! Name your five senses.

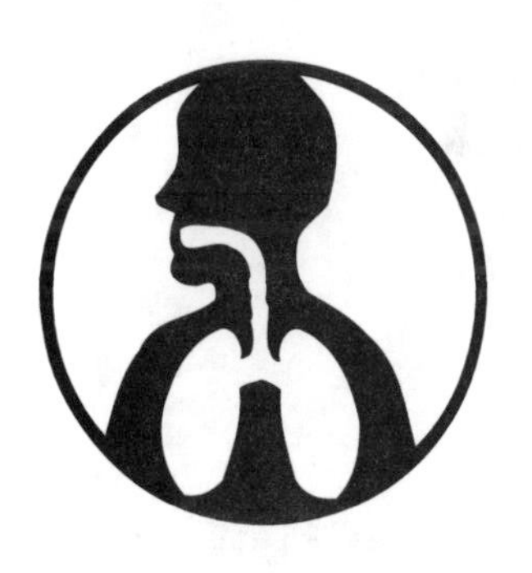

What is the colored part of your eye called?

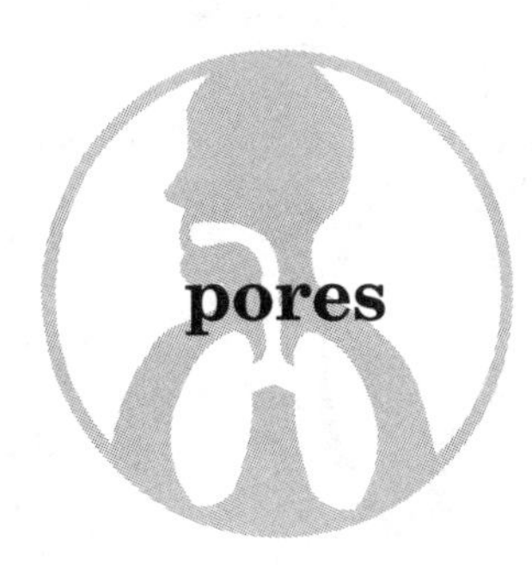

Human Body

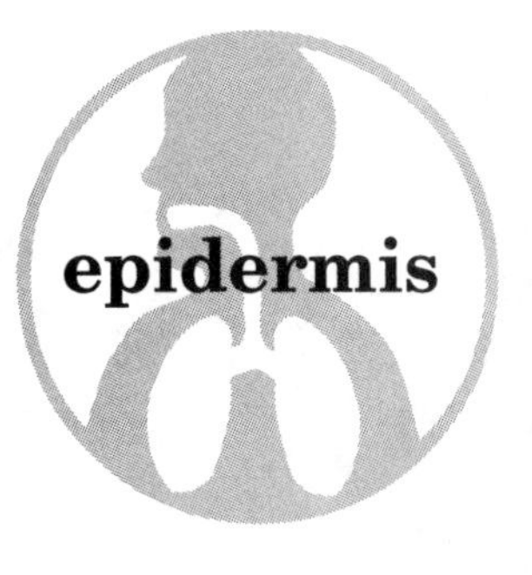

Human Body

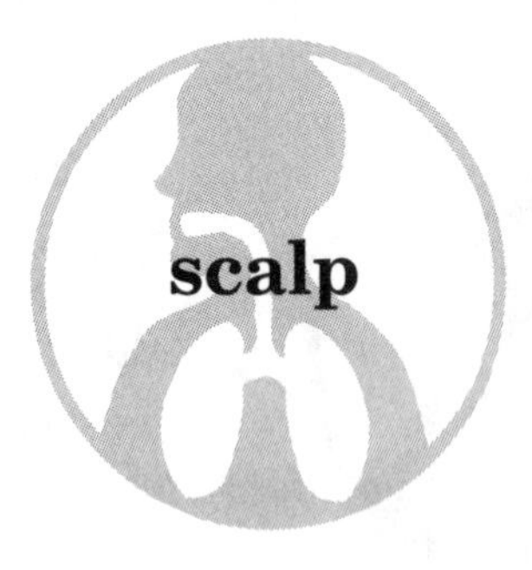

Human Body

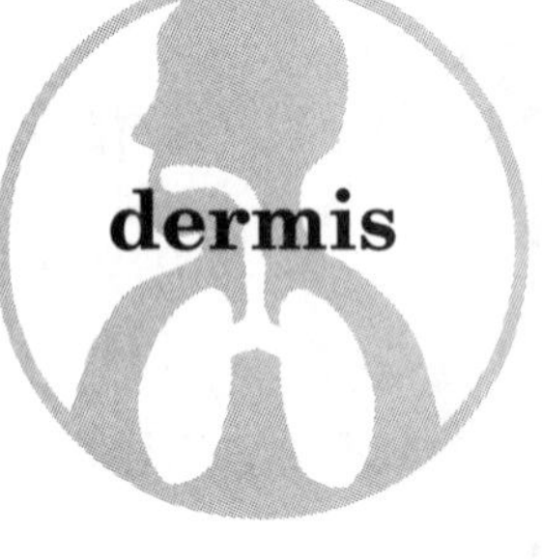

Human Body

Human Body

Human Body

Human Body

Human Body

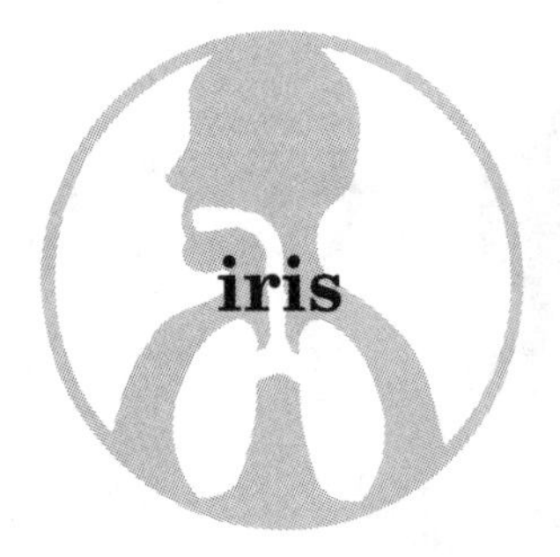

Human Body

sight, hearing,
smell, taste
and touch

Human Body

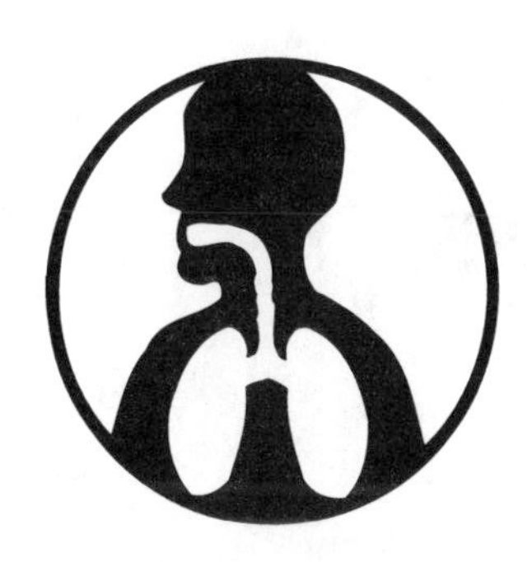

The lining at the back of the eye is called the _____.

This round opening in the middle of the eye lets light in.

True or False? You blink your eyes thousands of times a day.

The lens of the eye is located just behind the _____.

How do eyelashes help your eyes?

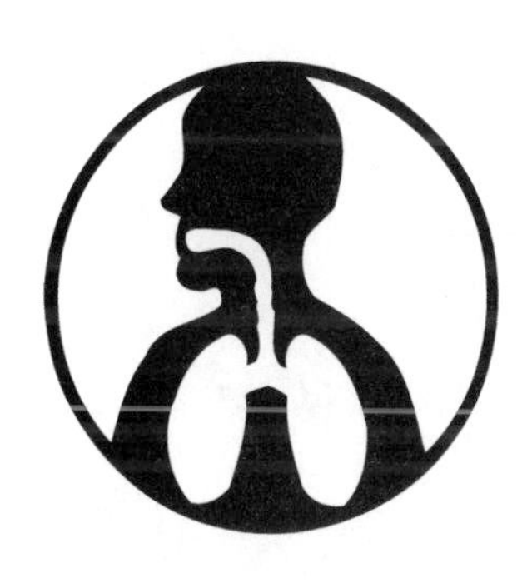

When it's dark, do the pupils of your eyes get bigger or smaller?

In bright light, do the pupils of your eyes get bigger or smaller?

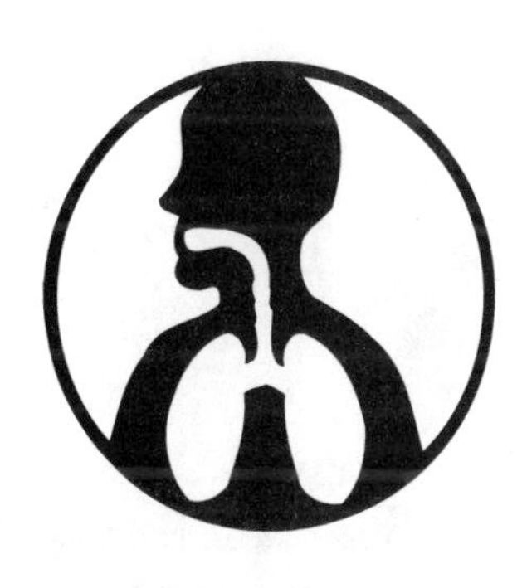

This tube carries sound from the outer part of the ear to the eardrum.

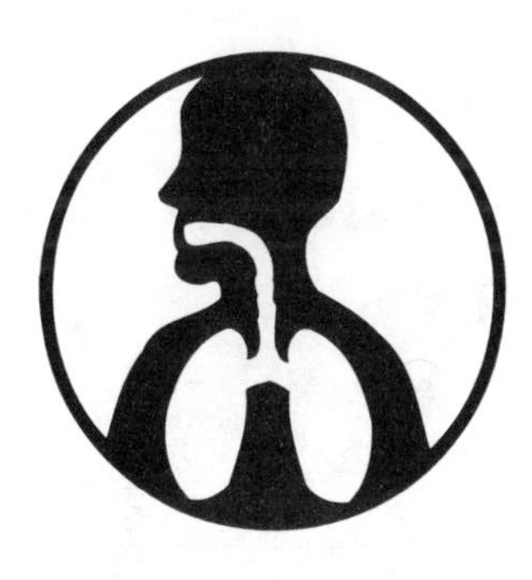

True or False? The smallest muscle in your body is located in your ear.

Sound causes the ear to _____.

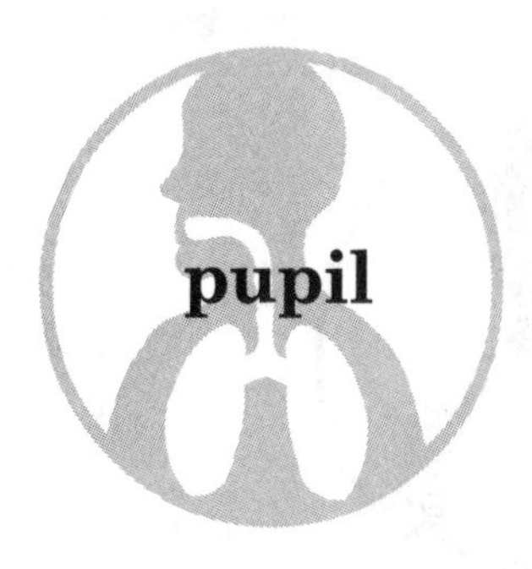

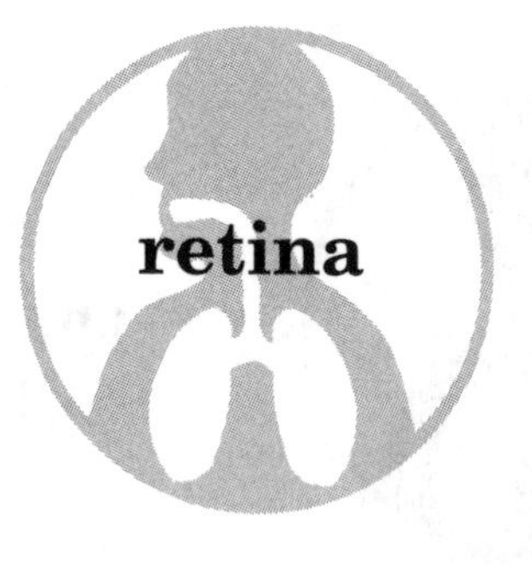

Human Body

Human Body

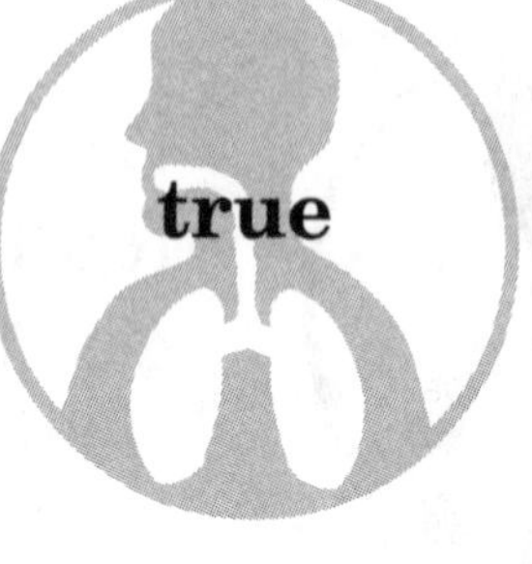

Human Body

Human Body

Human Body

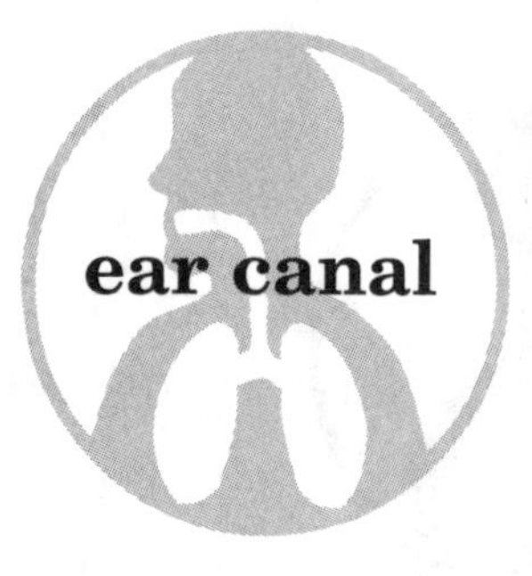

Human Body

smaller

Human Body

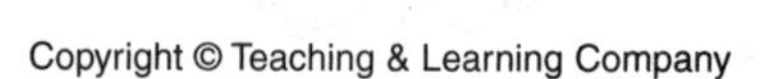

Human Body

True. It is called the stapes.

Human Body

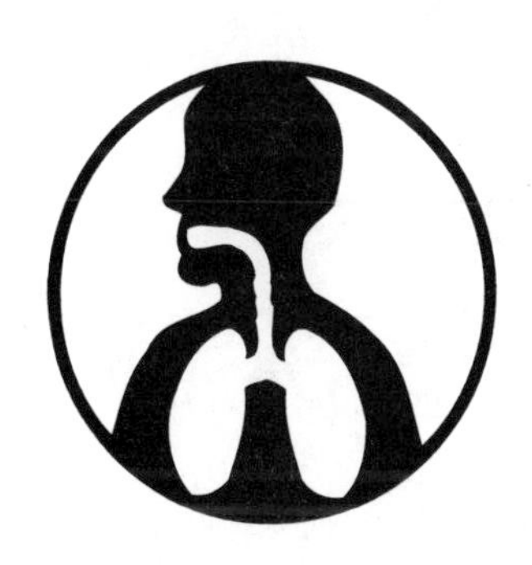

True or False? Soft sounds become more difficult to hear as people get older.

What are the two openings of the nose called?

True or False? Adults have a better sense of smell than children.

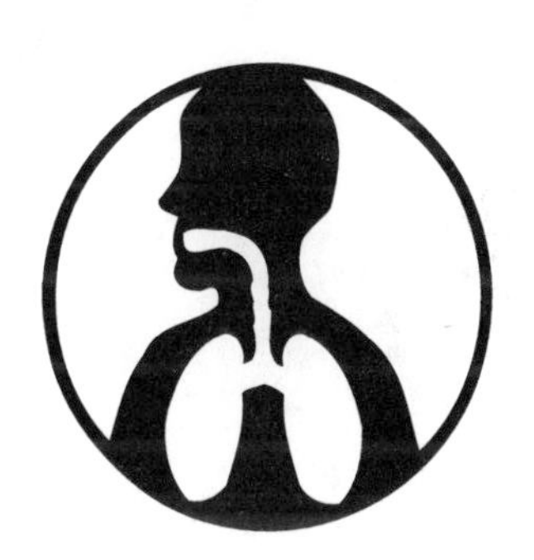

What are the little pink bumps on your tongue called?

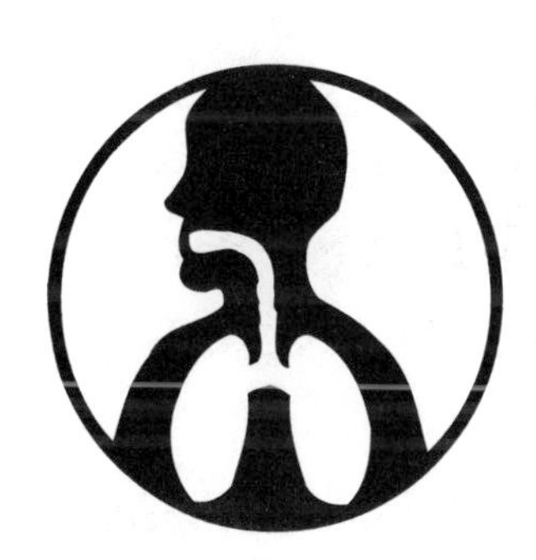

Do the taste buds on the tip of your tongue detect sweetness or sourness?

Is daily exercise important to good health?

When you exercise, you get more _____ into your lungs with each breath.

A pediatrician is a doctor who takes care of _____.

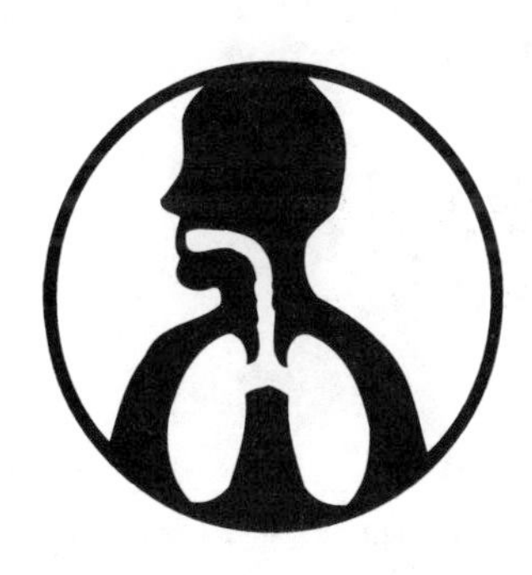

What is a dentist who straightens teeth called?

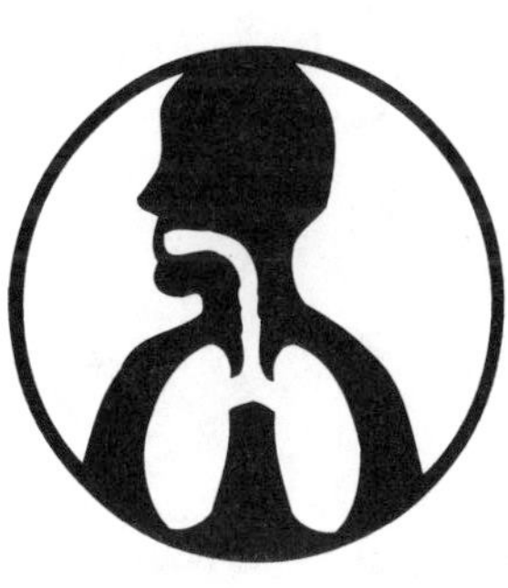

This ancient Greek doctor, often called “the father of medicine,” realized that diseases have natural causes.

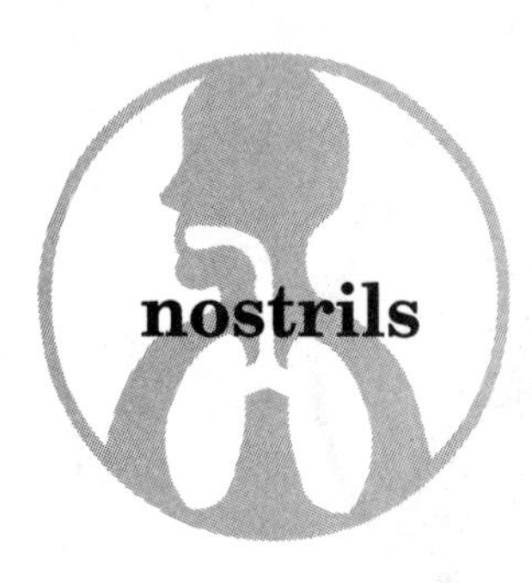

nostrils

Human Body

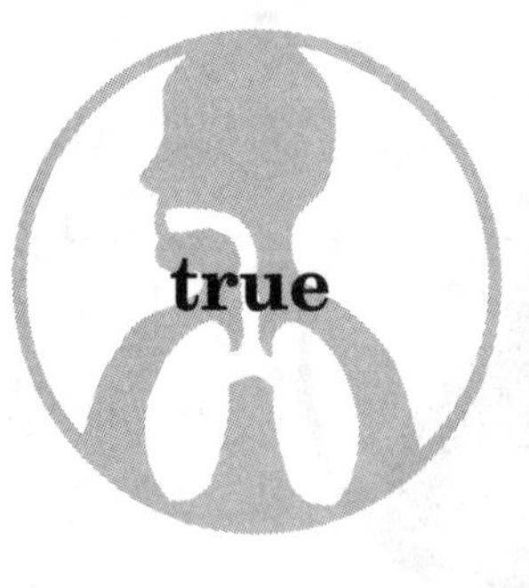

true

Human Body

taste buds

Human Body

False. Children have a better sense of smell.

Human Body

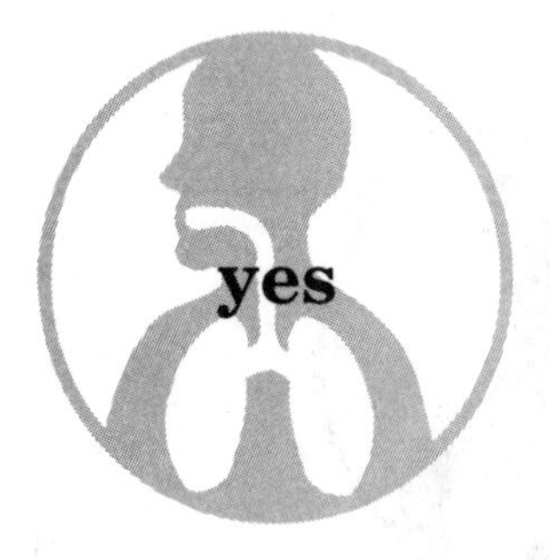

yes

Human Body

sweetness

Human Body

children

Human Body

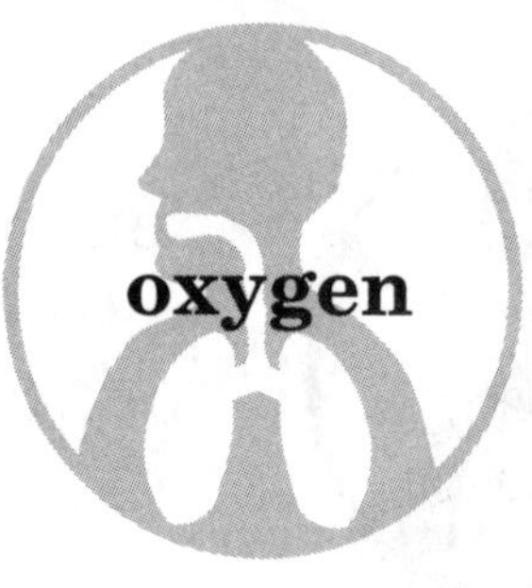

oxygen

Human Body

Hippocrates

Human Body

orthodontist

Human Body

A person who specializes in the study of insects is called an _____.

Insects have how many legs?

What is the hard outer covering of an insect called?

Are insects warm-blooded or cold-blooded animals?

Name the three main parts of an insect.

The legs and the wings are attached to what part of the insect?

What is the middle section of an insect's body called?

Insects breathe through tiny holes called _____.

What is the rear section of an insect's body called?

What is the silky covering for a pupa called?

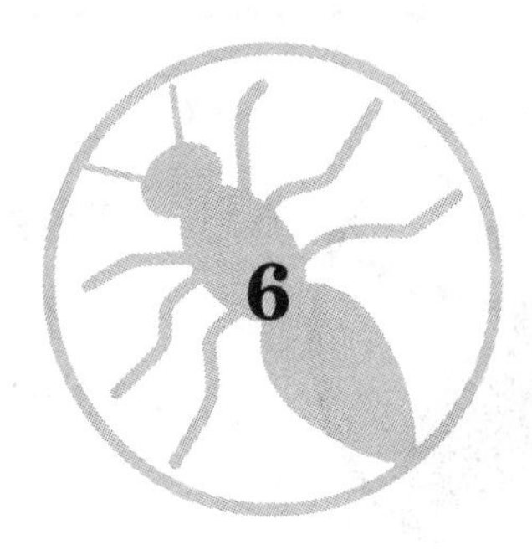

Insects & Arachnids

Insects & Arachnids

Insects & Arachnids

Insects & Arachnids

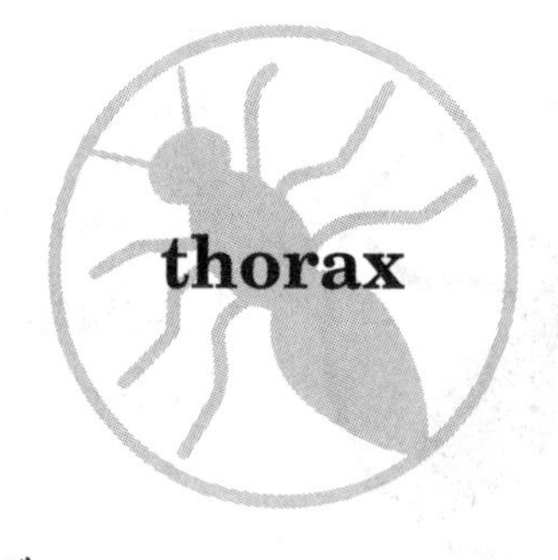

Insects & Arachnids

Insects & Arachnids

Insects & Arachnids

Insects & Arachnids

Insects & Arachnids

abdomen

Insects & Arachnids

What is the larva of a butterfly called?

What are young dragonflies called?

Name the process in which a larva will change into an adult insect.

Does the dragonfly have double wings?

What color is an insect's blood?

True or False? Dragonflies have excellent vision.

What are two reasons insects sting?

What is the name of the smallest fly in the world?

Ready, set, go! The fastest insect in the world is the _____. It can fly up to 60 mph.

A housefly's wings beat about how many times a second—50 or 200 times?

nymphs

Insects & Arachnids

caterpillar

Insects & Arachnids

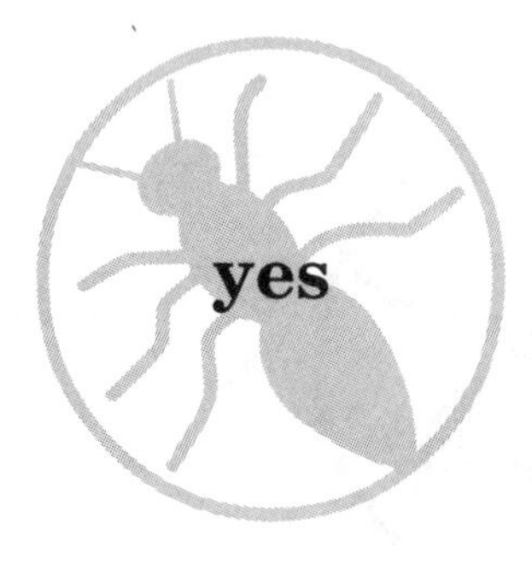

yes

Insects & Arachnids

metamorphosis

Insects & Arachnids

True. They can detect movement up to 40 feet away.

Insects & Arachnids

yellow

Insects & Arachnids

midge

Insects & Arachnids

to defend themselves and to catch prey

Insects & Arachnids

200 times

Insects & Arachnids

dragonfly

Insects & Arachnids

What is the longest insect in the world?

What colors are grasshoppers?

What happens if a walkingstick loses a leg?

What does a grasshopper do when threatened?

World Champion! Name the loudest insect in the world.

Beware! This insect "attacks" wood.

True or False? Only the female crickets chirp.

What insect builds the biggest nest?

In some parts of the world, crickets are considered a sign of _____.

What insect is sometimes called the "armored tank" of the insect world?

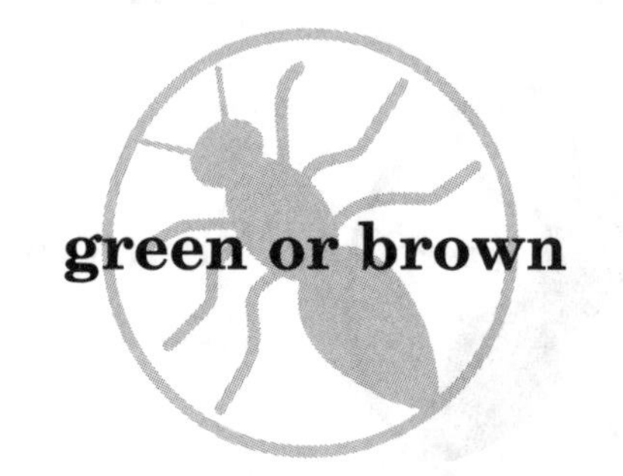

green or brown

Insects & Arachnids

giant stick insect of Indonesia

Insects & Arachnids

It spits a brown liquid.

Insects & Arachnids

It grows another one.

Insects & Arachnids

termite

Insects & Arachnids

cicada

Insects & Arachnids

termite

Insects & Arachnids

False. Only the male crickets chirp.

Insects & Arachnids

beetle

Insects & Arachnids

good luck

Insects & Arachnids

Fireflies are not flies, they are _____.

Pill bug is the nickname for what bug?

A firefly's "light" is produced by a special _____.

Rolypolies belong to a special group of animals called _____.

Watch out! These bugs give off a bad odor when disturbed.

What do ladybugs like to eat?

It's a world record! Name the heaviest insect in the world.

What are the hard wing cases of a ladybug called?

These beetles can shoot a hot liquid from their abdomens.

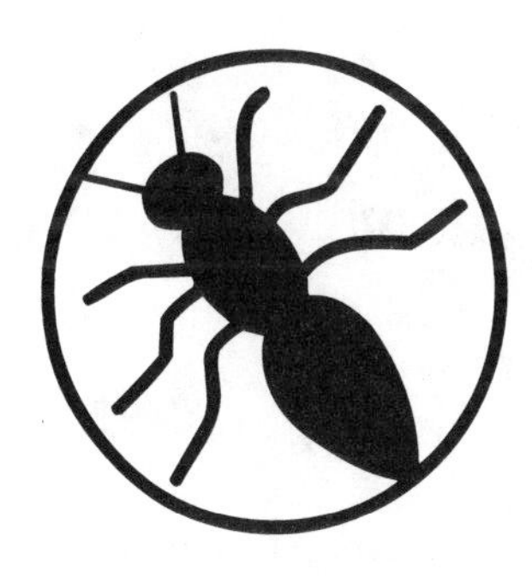

Butterflies and moths belong to a group called _____.

rolypoly

Insects & Arachnids

beetles

Insects & Arachnids

isopods

Insects & Arachnids

chemical

Insects & Arachnids

aphids

Insects & Arachnids

stink bug

Insects & Arachnids

elytra

Insects & Arachnids

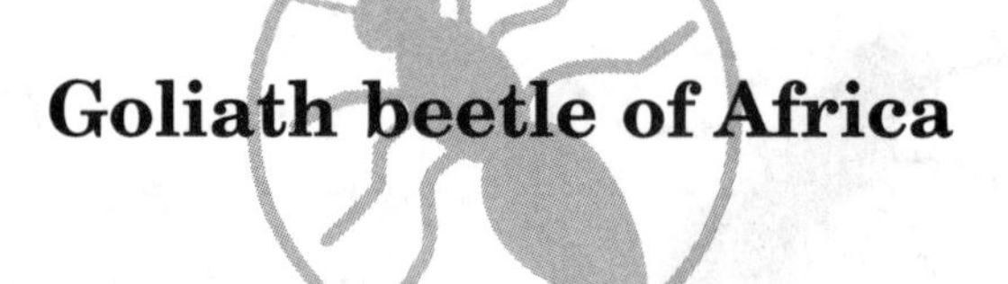

Goliath beetle of Africa

Insects & Arachnids

lepidoptera

Insects & Arachnids

bombardier beetles

Insects & Arachnids

Butterflies have a tube that is used to drink nectar. What is this tube called?

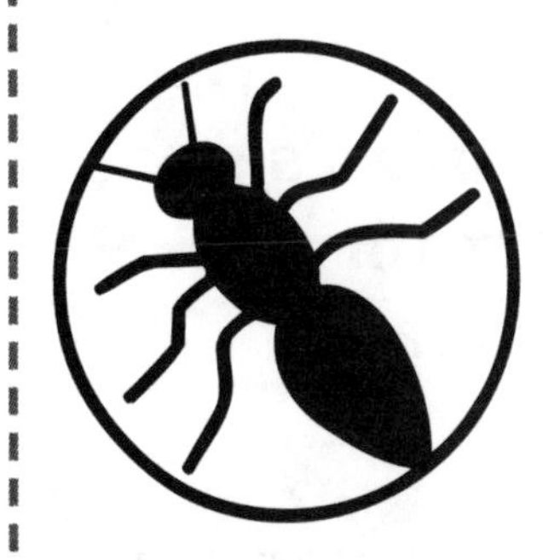

What is the name of the largest butterfly in the world? It has a wingspan of about 11 inches!

How do butterflies help plants and flowers grow?

Name the smallest butterfly in the world.

Butterfly and moth wings are covered by tiny _____.

What butterflies hold the record for the longest migration? They migrate over 2000 miles in the fall.

True or False? Most butterflies rest with their wings folded up.

What butterflies have see-through wings that make them hard to spot?

Butterflies migrate to _____ places during wintertime.

Name the three main parts of a caterpillar.

Queen Alexandra's birdwing from New Guinea

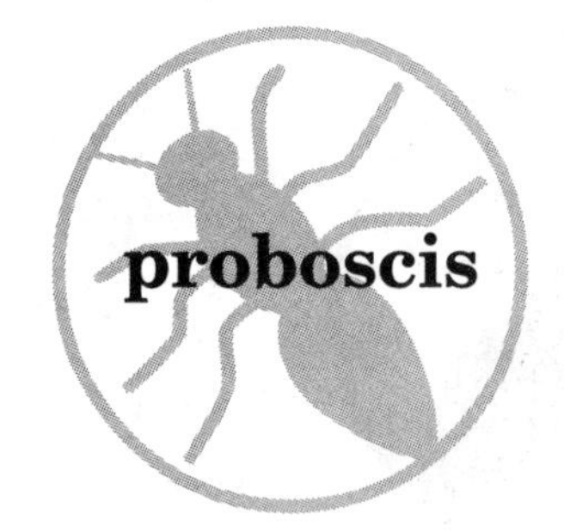

proboscis

western pygmy blue

by carrying pollen from one flower to another

monarch butterflies

scales

glasswing butterflies

true

head, thorax and abdomen

warm

Most caterpillars change their skin about how many times?

True or False? Most moths have knobs on the ends of their antennae.

What two seasons are the best times to look for caterpillars?

Do moths tend to fly at night or during the day?

Do caterpillars have a big appetite?

What moth has the same name as a bird of prey?

There are about how many species of moths in the world–50,000 or 100,000?

Ants live in groups called ______.

What is the name of the world's biggest moth?

True or False? A colony of leaf-cutter ants can have millions of ants.

Insects & Arachnids

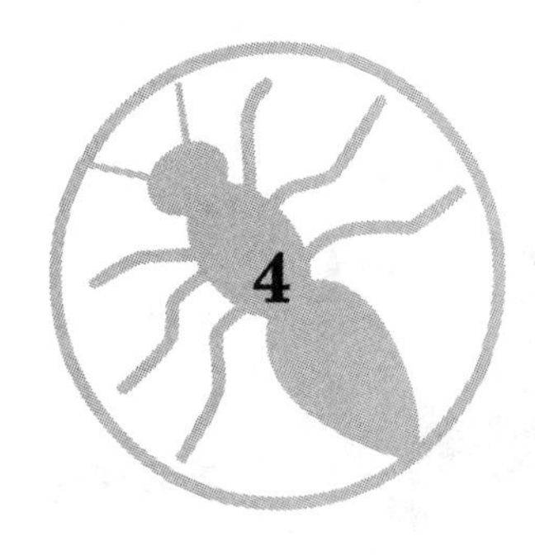

Insects & Arachnids

Insects & Arachnids

Insects & Arachnids

Insects & Arachnids

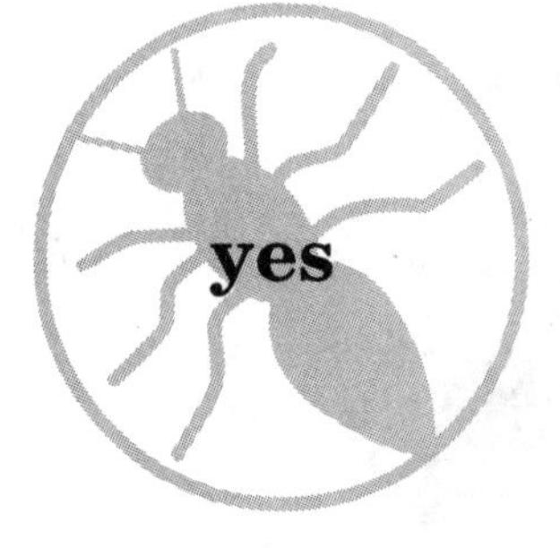

Insects & Arachnids

Insects & Arachnids

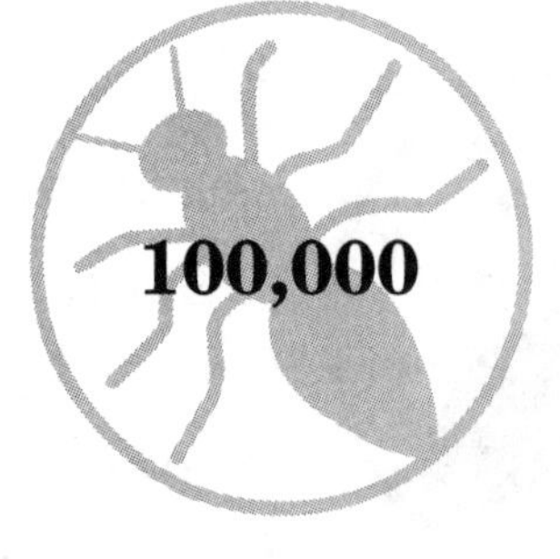

Insects & Arachnids

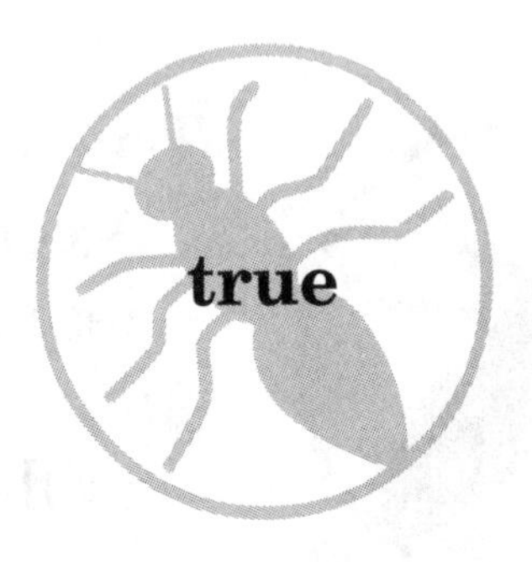

Insects & Arachnids

atlas moth

Insects & Arachnids

What is the pouch inside an ant's abdomen called?

A male bee is called a _____.

What is a queen ant's main job?

Bees live in groups called _____.

Where do harvester ants live?

Bees make honey from _____.

True or False? Most ant species make their nests underground.

Bees make their honeycombs out of wax. Each honeycomb cell has how many sides?

Name the three types of bees.

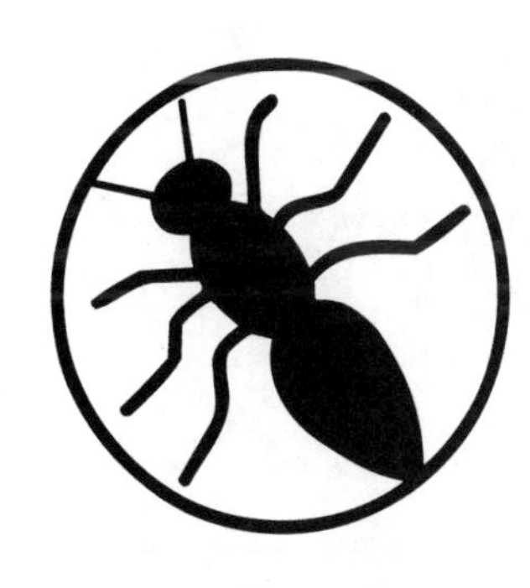

Who is the leader of a bee colony and lays the eggs?

Insects & Arachnids

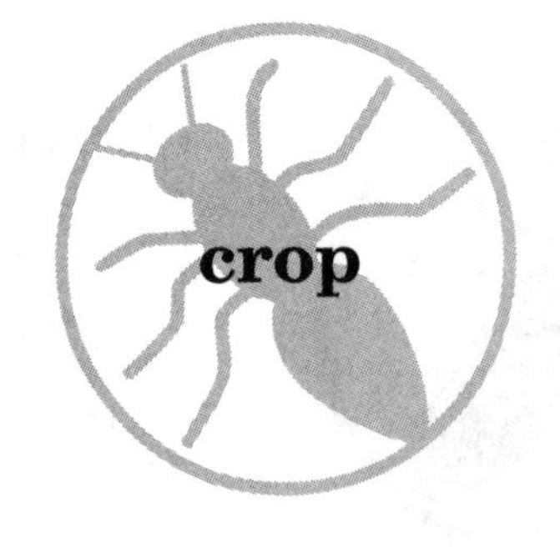

Insects & Arachnids

Insects & Arachnids

Insects & Arachnids

Insects & Arachnids

Insects & Arachnids

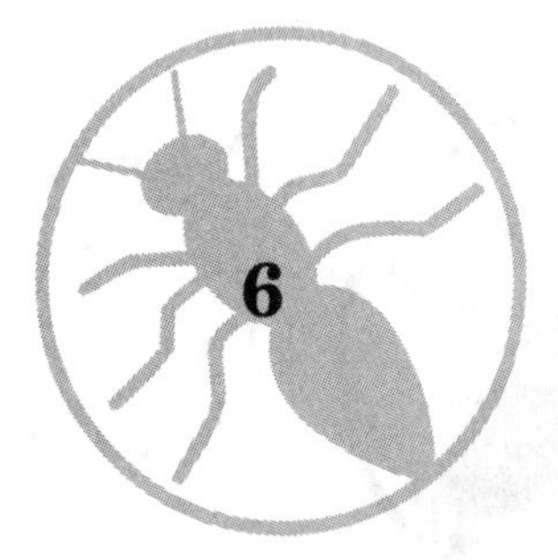

Insects & Arachnids

Insects & Arachnids

Insects & Arachnids

queen, workers
and drones

Insects & Arachnids

A bee smells with a pair of long feelers called _____.

What type of wasp builds the largest nest?

What happens when an older hive gets too large?

Spiders belong to an animal group called _____.

Where do mining bees dig their tunnels?

True or False? There are more than 30,000 spider species on Earth.

How do bees communicate with each other?

Name the two main parts of a spider.

A bee has a "basket" on its back legs to hold _____.

How many legs does a spider have?

antennae

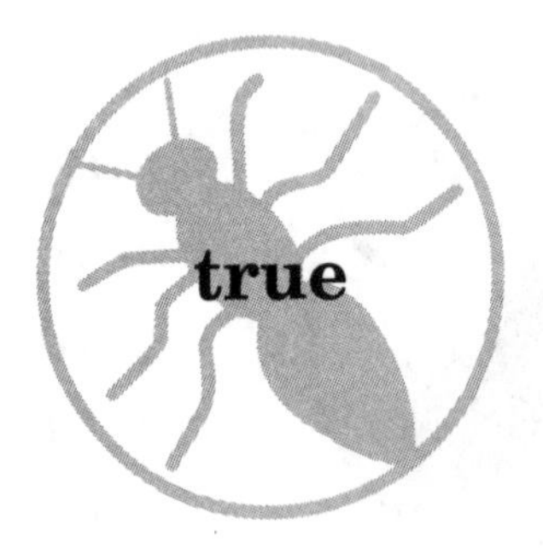

in the ground

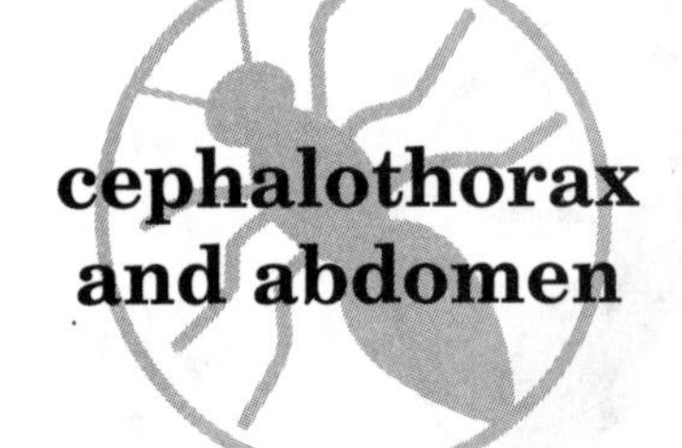

by dancing

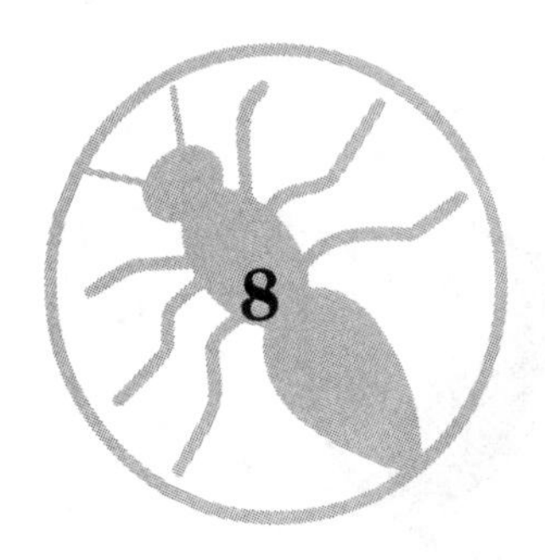

pollen

A baby spider is called a _____.

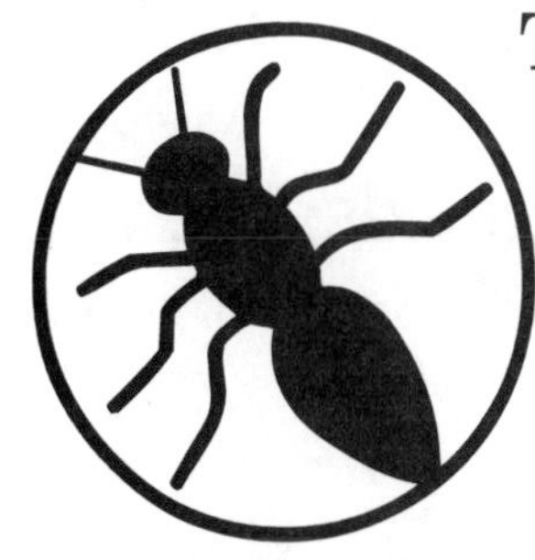

Threads of a spider-web that spread straight outward from the center of the web are called _____ threads.

What are the silk-spinning organs of a spider called?

Name two enemies of spiders.

If a spider loses a leg, can it grow a new one?

Ready, set, jump! These hunting spiders jump on their prey.

A spider can have as many as _____ eyes.

What spider has the same name as a crustacean?

True or False? As a spider gets bigger, it sheds its skin.

Knock, knock! Where do trapdoor spiders live?

radial threads

Insects & Arachnids

spiderling

Insects & Arachnids

Answers will vary. Some may include birds, wasps and other spiders and small animals.

Insects & Arachnids

spinnerets

Insects & Arachnids

jumping spiders

Insects & Arachnids

yes

Insects & Arachnids

crab spider

Insects & Arachnids

8

Insects & Arachnids

in underground burrows

Insects & Arachnids

true

Insects & Arachnids

Are trapdoor spiders mainly found in cold or tropical regions?

Most parts of a tarantula's body are covered with _____.

What spider has the same name as a mammal that resembles a dog?

Does a scorpion hunt during the daytime or at night?

This spider spits a net of poison and "glue" to capture insects.

Are tarantulas herbivores or carnivores?

Does a swamp spider live on the surface of the water or by the water's edge?

A tarantula's _____ can paralyze its prey.

Water spiders trap bubbles of _____ in their underwater webs.

Do female tarantulas live longer than the males?

hair

Insects & Arachnids

tropical regions

Insects & Arachnids

at night

Insects & Arachnids

wolf spider

Insects & Arachnids

carnivores

Insects & Arachnids

spitting spider

Insects & Arachnids

venom

Insects & Arachnids

on the surface
of the water

Insects & Arachnids

Yes. Some female
tarantulas have
lived up to 20 years!

Insects & Arachnids

air

Insects & Arachnids

What is a person who studies the science of the oceans called?

Tides are caused by the pull of the moon's _____ on the oceans.

It's a world record! Name the biggest ocean in the world.

Parts of the ocean flow like rivers. What are these moving areas called?

True or False? The Pacific Ocean is one-third of Earth's surface.

What is the top part of an ocean wave called?

What is the name of the second largest ocean in the world?

What is the bottom of the ocean called?

The Atlantic Ocean is bordered by what two continents on the west?

When is it a good time to find empty seashells on the beach—during high tide or low tide?

gravity

Oceanography

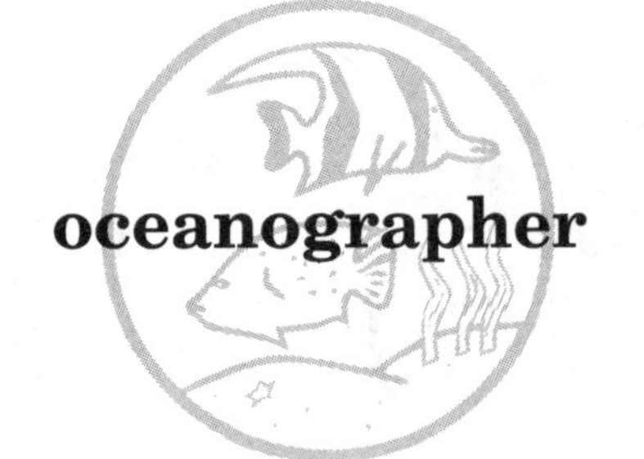

oceanographer

Oceanography

ocean currents

Oceanography

Pacific Ocean

Oceanography

crest

Oceanography

true

Oceanography

ocean floor

Oceanography

Atlantic Ocean

Oceanography

low tide

Oceanography 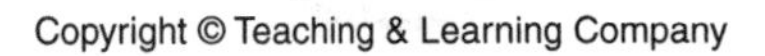

North America and
South America

Oceanography

The land that slants down from the shore into the ocean is called the continental _____.

The Indian Ocean is bordered by what continent on the west?

A piece of land surrounded by water is called an _____.

True or False? Much of the Indian Ocean is located within the tropics.

These giant waves can swamp whole islands.

A narrow waterway connecting two seas is called a _____.

What is the third largest ocean on Earth? It's less than half the size of the Pacific Ocean.

What is the name of the smallest ocean on Earth?

Most of the Indian Ocean lies in what hemisphere—Northern or Southern?

Most of the Arctic Ocean is within what imaginary line?

Africa

shelf

Oceanography

Oceanography

true

island

Oceanography

Oceanography

strait

tsunami

Oceanography

Oceanography

Arctic Ocean

Indian Ocean

Oceanography

Oceanography

Arctic Circle

Southern Hemisphere

Oceanography

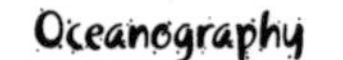

Oceanography

What strait links the Arctic Ocean to the Pacific Ocean?

What kind of coral looks like a big flat dinner dish?

A large mass of ice that is floating in the sea is called an _____.

What sea separates Africa from Europe?

The world's largest coral reef is located in Australia. Name this well-known tourist attraction.

World Champion Whale! What is the name of the largest whale in the world?

Coral reefs are made from the _____ of billions of tiny animals called coral polyps.

True or False? A baby blue whale at birth weighs more than a full-grown elephant.

Staghorn coral looks like deer _____.

Whales without teeth, such as the blue whale, are known as _____ whales.

plate coral

Oceanography

Bering Strait

Oceanography

Mediterranean Sea

Oceanography

iceberg

Oceanography

blue whale

Oceanography

Great Barrier Reef

Oceanography

true

Oceanography

skeletons

Oceanography

baleen

Oceanography

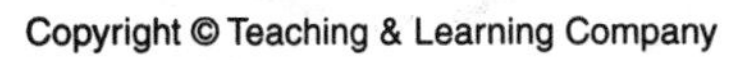

antlers

Oceanography

What whale has the biggest tongue? Its tongue can weigh over one ton.

The opening on the top of a whale used for taking in air is called a _____.

Whales travel together in groups called _____.

These small shellfish attach themselves to whales, ships and other things.

What whales make the longest migration of any mammal?

Baby whales are called _____.

Are whales cold-blooded or warm-blooded?

This light-colored, toothed whale has the nickname "sea canary."

What are the whale's wide, muscular tail fins called?

What whale is sometimes called the "unicorn whale"?

blowhole

Oceanography

bowhead whale

Oceanography

barnacles

Oceanography

pods

Oceanography

calves

Oceanography

gray whales

Oceanography

beluga whale

Oceanography

warm-blooded

Oceanography

narwhal

Oceanography

flukes

Oceanography

True or False?
A long spiral
tusk is only
found on the
female narwhal.

World Champion!
Name the world's
largest fish. They
can grow to over
60 feet long.

Narwhals and
beluga whales
are found in
what part of
the world?

What is the
name of the
world's smallest
shark?

Killer
whales are
also called
______.

What yellow-
colored shark
has the name
of a fruit?

What is the
name of the
largest seal in
the world? It can
weigh as much
as 8000 pounds.

This big-
mouthed shark
was discovered
in 1976 off the
coast of Hawaii.

A walrus uses
its ______ as
hooks when
climbing onto
the ice.

What shark
has the
name of an
Asian cat?

whale shark

Oceanography

False. Tusks are only found on the male narwhal.

Oceanography

dwarf dogfish

Oceanography

The Arctic

Oceanography

lemon shark

Oceanography

orcas

Oceanography

megamouth

Oceanography

northern elephant seal

Oceanography

tiger shark

Oceanography

tusks

Oceanography

Splash!
What shark
can swim
the fastest?

What happens
when a
porcupine
fish is
frightened?

A shark's
skeleton is
made up
of _____.

This fish can
sometimes leave
the water with
enough height to
enable it to land
on a ship's deck.

A shark
breathes
through its
_____.

What fish is
nicknamed
the "tiger of
the sea"?

Zoom!
What is the
fastest fish
in the
world?

Is tuna found
in the colder or
warmer oceans
of the world?

Watch out!
What fish
sometimes
damages boats
with its
sharp nose?

What is the
main natural
enemy of tuna?

It gulps water and inflates its body to look bigger.

mako shark

flying fish

cartilage

barracuda

gills

warmer

sailfish

killer whale

swordfish

What fish is sometimes called a stingaree?

How did the parrotfish get its name?

Where does a stingray usually live?

How many arms does an octopus have?

A stingray's _____ can cause serious wounds.

If threatened, what can an octopus do?

What fish is also known as the devil ray or devil fish?

Do octopuses live long or short lives?

Clown fish swim among the stinging _____ of the sea anemone.

A squid has how many fins at the tail end?

Its mouth looks like a parrot's beak.

stingray

8

partially buried in mud or sand

It can release a cloud of inky fluid.

spines

short lives

manta ray

2

tentacles

True or False?
A squid can
only move
forward in
the water.

True or False?
The sea horse
is a good
swimmer.

What is the name
of the largest squid
in the world? They
can measure up to
66 feet in length.

Where does
a sea horse
live?

Jellyfish have
stinging cells
on their
______.

Can a sea
horse
change
colors?

Is a jellyfish
a vertebrate
or an
invertebrate?

True or False?
A sea star has
a mouth on the
top of its body.

This fish's
name means
"bent horse"
in Latin.

Most sea
stars have
how many
arms?

false

False. It can move both forward and backward.

Oceanography

Oceanography

in tall ocean weeds

giant squid

Oceanography

Oceanography

yes

tentacles

Oceanography

Oceanography

False. It is located on the bottom of its body.

invertebrate

Oceanography

Oceanography

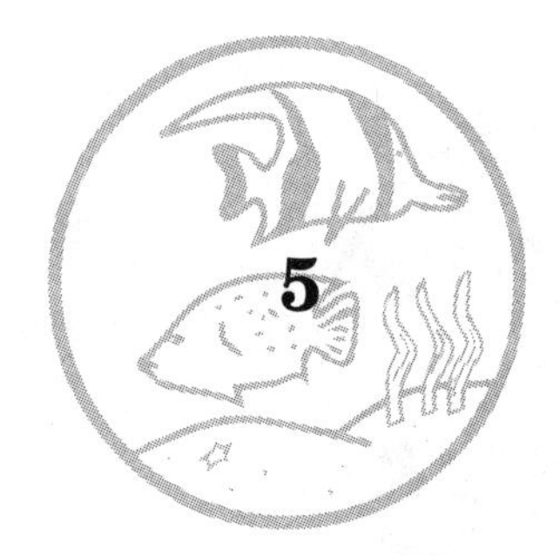

5

sea horse

Oceanography

Oceanography

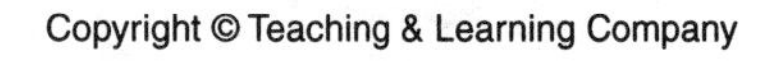

What are the colors of sea stars?

A hermit crab looks for an empty _____ for its home.

Crabs, lobsters and shrimp belong to a group of animals called _____.

A lobster has _____ pairs of legs.

Name the three main parts of an adult crustacean.

It's a record-setter! What sea turtle is the largest of all turtles? It can weigh as much as 1200 pounds!

The hard shell of a crab that covers its head and thorax is called the _____.

Do sea turtles lay their eggs in the water or on land?

True or False? The Japanese spider crab is the biggest crab in the world.

True or False? A sea turtle's shell is lighter than those of a land turtle.

Oceanography

red, orange, purple, blue and brown

Oceanography

Oceanography

crustaceans

Oceanography

Oceanography

head, thorax, and abdomen

Oceanography

Oceanography

carapace

Oceanography

Oceanography

true

Oceanography

Migration champion! What sea bird travels farther than any other bird?

Thick layers of _____ keep penguins warm in cold water.

A pelican has a long beak with an attached _____ that makes a handy "scoop" for fish.

Can penguins fly?

True or False? Pelicans usually travel in flocks.

What is the name of the largest penguin in the world?

Do penguins live south or north of the equator?

Name the smallest penguin in the world.

A place where large numbers of penguins live is called a _____.

What are large, deep cracks on the ocean floor called?

blubber

Oceanography

Arctic tern

Oceanography

no

Oceanography

pouch

Oceanography

emperor penguin

Oceanography

true

Oceanography

fairy penguin

Oceanography

south

Oceanography

trenches

Oceanography

colony or rookery

Oceanography

A galaxy is a group of billions of _____.

Earth rotates on its axis every _____ hours.

The sun, Earth, and planets are part of what galaxy?

Name the four "inner planets" of our solar system.

What is the shape of the Milky Way—spiral or elliptical?

Hot zone! What is the hottest planet in our solar system?

Planets circle the sun in paths called _____.

The coldest of all the known planets is _____.

The imaginary line running through the center of a planet is called an _____.

What planet is closet to the sun?

Solar System

stars

Solar System

Mercury, Venus, Earth and Mars

Solar System

Milky Way

Solar System

Venus

Solar System

spiral

Solar System

Solar System

orbits

Solar System

Solar System

Axis

Solar System

How many days does it take Mercury to orbit the sun–88 or 100 days?

Earth is the _____ largest planet in our solar system.

What planet is often called "Earth's twin" because the two planets are very close in size?

It takes Earth about _____ days to orbit the sun.

The planet Venus was named after the Roman goddess of _____.

Earth is surrounded by a thick layer of gases called the _____.

This planet is nicknamed the "Evening Star."

What is the force that pulls objects toward the center of the earth?

Name the third planet from the sun.

Name the British scientist who first explained how gravity works.

fifth

Solar System

88 Earth days

Solar System

about 365 days

Solar System

Venus

Solar System

atmosphere

Solar System

love

Solar System

gravity

Solar System

Venus

Solar System

Sir Isaac Newton

Solar System

Earth

Solar System

What is the center of the earth called?

The highest known mountain in our solar system is located on Mars. Name it.

The thin surface (top) layer of Earth that we live on is called the _____.

Mars has the biggest canyon in our solar system. Name it.

What planet is nicknamed "The Red Planet"?

The planet Mars was named after what Roman god?

True or False? Mars is about half the size of Earth.

What is the name of the largest planet in our solar system?

What causes the reddish-color of Mars?

Jupiter is the _____ planet from the sun.

Olympus Mons

Solar System

inner core

Solar System

Mariner Valley

Solar System

crust

Solar System

Roman god of war

Solar System

Mars

Solar System

Jupiter

Solar System

true

Solar System

fifth

Solar System

the rust in its iron-rich soil

Solar System

What is the Great Red Spot on Jupiter?

Name the second-largest planet in our solar system?

Ganymede is the largest moon on what planet?

What are Saturn's rings made of?

True or False? If Jupiter were a hollow ball, over 1000 Earths could fit inside it.

What is the name of Saturn's largest moon?

Jupiter's magnetic field is how many times stronger than Earth's— 5 or 10 times?

The planet Saturn is named after the Roman god of _____.

What Italian astronomer was the first person to look at Jupiter through a telescope?

True or False? Neptune's rings are darker and thinner than Saturn's rings.

Saturn

Solar System

a gigantic storm

Solar System

ice, rock and dust

Solar System

Jupiter

Solar System

Titan

Solar System

true

Solar System

agriculture

Solar System

10 times

Solar System

true

Solar System

Galileo Galilei

Solar System

This blue-green planet is the seventh planet in our solar system.

What kind of gas in Neptune's atmosphere makes the planet look blue?

The planet Uranus was named after the Roman god of the _____.

What is the largest moon on Neptune—Proteus or Triton?

Which planet was the first planet to be discovered with a telescope—Uranus or Neptune?

What is the smallest planet in our solar system?

Name the eighth planet from the sun.

An American astronomer named Clyde Tombaugh discovered the dwarf planet Pluto in what year—1930 or 1950?

The planet Neptune was named after what Roman god?

How many years does it take Pluto to orbit the sun—about 248 years or 300 years?

methane

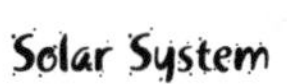

Uranus

Triton

Solar System

sky

Solar System

Pluto

Solar System

Uranus

Solar System

1930

Solar System

Neptune

Solar System

about 248 years

Solar System

Roman god of the sea

Solar System

The sun is not a planet. It's a _____.

What are dark patches on the sun's surface called?

Why does the sun look bigger than other stars?

Are sunspots cooler or hotter than the gases around them?

How many minutes does it take light from the sun to reach Earth?

The center of the sun is called the _____.

The layer of thin gas all around the sun is called the _____.

Does sunlight have a great influence on Earth's weather?

True or False? The sun's energy is produced at its surface.

Where is the hottest part of the sun located?

sunspots

Solar System

star

Solar System

cooler

Solar System

because it is closer to Earth

Solar System

core

Solar System

8

Solar System

yes

Solar System

corona

Solar System

center or core

Solar System

False. At its center

Solar System

What happens during a solar eclipse?

What are the patterns of stars in the sky called?

A person who studies the stars, planets and space is called an _____.

How many constellations do astronomers recognize?

A building with telescopes where astronomers study the skies is called an _____.

What is the name of the star at the end of the Little Dipper's handle?

What type of telescope lets people look at a reflection of stars by using two mirrors?

This constellation was named after a famous hunter because people thought the group of stars looked like a hunter with a club and shield.

People go to this building where a special projector shows the images of the sky on a domed ceiling.

What constellation was named after a magic flying horse in an ancient story?

constellations

Solar System

The moon comes between the sun and Earth.

Solar System

88

Solar System

astronomer

Solar System

North Star

Solar System

observatory

Solar System

Orion

Solar System

reflecting telescope

Solar System

Pegasus

Solar System

planetarium

Solar System

Binary stars are pairs of stars held together by _____.

What is a light-year?

What is a nebula?

The surface of the moon is covered with deep _____.

True or False? The coolest stars glow the reddest.

What caused the craters on the moon?

At the end of their lives stars become white _____.

The sky is always what color on the moon?

What is the final explosion of a dying supergiant star called?

Name the Apollo astronaut who became the first man to set foot on the moon.

the distance light travels in one year

gravity

craters

a cloud of gas and dust in space

space rocks crashing into the moon's surface

true

black

dwarfs

Neil Armstrong

supernova

When our moon shines, it is _____ light from the sun.

Nifty nickname! What is the nickname for a meteor?

A day on the moon lasts one Earth _____.

Asteroids are pieces of rock that orbit the _____.

Frozen balls of ice, gas and rock that orbit the sun are called _____.

Where is the asteroid belt located?

The tail of a comet points away from the _____.

What are people who travel in space called?

True or False? The tail of a comet may be millions of miles long.

It's a world record! A Russian satellite was the first man-made satellite. Name it.

shooting star

Solar System

reflecting

Solar System

sun

Solar System

month

Solar System

between Mars and Jupiter

Solar System

comets

Solar System

astronauts

Solar System

sun

Solar System

Sputnik

Solar System

true

Solar System

What animal was the first living being to orbit Earth in 1957?

A place where rockets take off is called a launch _____.

What is the name of the first human to travel in space in 1961?

A small rocket that is attached to a large space rocket is called a _____ rocket.

Name the first American astronaut to orbit in space in 1962.

What do you call an unmanned spacecraft that is controlled completely from Earth?

What is the name of the first American woman in space?

What does *UFO* mean?

What do you call the part of a space shuttle where the equipment is stored?

The Hubble Space Telescope was sent into orbit in what year—1960 or 1990?

pad

Solar System

a dog named Laika

Solar System

booster

Solar System

Yuri Gagarin

Solar System

probe

Solar System

John H. Glenn, Jr.

Solar System

Unidentified Flying Object

Solar System

Sally Ride

Solar System

1990

Solar System

cargo bay

Solar System

Scientists who study weather and climate are called _____.

What is the push of air on the earth called?

What do we call the study of weather?

What does a barometer measure?

What weather instrument measures air temperature?

True or False? The first barometer was made by an Italian man named Evangelista Torricelli in 1643.

What is the name of the man who invented the first thermometer?

Moving air is called _____.

True or False? The first thermometer was invented in 1800.

A flow of air or water is called a _____.

Weather

Weather

air pressure

Weather

meteorology

Weather

true

Weather

thermometer

Weather

Weather

Gabriel Daniel
Fahrenheit

Weather

current

Weather

False. The first
Fahrenheit thermometer
was invented in 1714.

Weather

What does a weather vane indicate?

The kind of weather that a region has over a long time is called its _____.

Some windmills are used to turn wind power into _____.

The water on Earth travels on an "endless journey" called the water _____.

An anemometer measures wind _____.

What happens when water turns to water vapor?

How many numbers are used in the Beaufort Scale to describe wind strength?

The amount of moisture in the air is called _____.

What takes pictures of Earth's weather from space?

What do hygrometers measure?

Weather

the direction of the wind

Weather

Weather

electricity

Weather

Weather

speed

Weather

Weather

12

Weather

Weather

satellites

Weather

True or False? Air is thickest nearer the ground.

These dark thick clouds can produce heavy rain, hail or thunderstorms.

Water vapor in the air forms ______.

What is the process called by which chemicals are sprayed into clouds in an attempt to cause rain?

What is any form of water falling through the sky called?

Is hot air lighter or heavier than cold air?

What low clouds are puffy and look like piles of fluffy white cotton?

The process of water turning from gas into a liquid is called ______.

High-level cirrus clouds are made up of ice ______.

Earth's atmosphere contains how many layers?

Weather

true

Weather

cloud seeding

Weather

clouds

Weather

Hot air is lighter than cold air.

Weather

precipitation

Weather

condensation

Weather

cumulus clouds

Weather

5

Weather

crystals

Weather

The first layer of Earth's atmosphere is called the _____.

Is mist thinner or thicker than fog?

Name the layer of the atmosphere that contains the ozone layer.

Fog forms when water vapor condenses near the _____.

Ozone is a type of _____.

When does a fogbow occur?

Name the highest layer of the atmosphere that merges into space.

Smog occurs when smoke and other kinds of pollutants mix with _____.

Most weather occurs in which layer of the atmosphere—the troposphere or exosphere?

A brief rainfall is called a _____.

Weather

troposphere

Weather

ground

Weather

stratosphere

Weather

when sunlight passes through a patch of fog

Weather

gas

Weather

fog

Weather

exosphere

Weather

shower

Weather

troposphere

Weather

How many colors does a rainbow have?

True or False? Rainy tropical islands and rain forests get rainbows less frequently.

Name the seven colors in a rainbow.

_____ is a big electrical spark that occurs during a thunderstorm.

True or False? Rainbows are always opposite the sun.

Lightning flashes with a crash of _____.

Rainbows are caused by sunlight passing through _____.

What is the name of a big dark cloud that brings heavy rain, thunder and lightning?

Rainbows make a shape called an _____.

What kind of storm usually causes flash floods?

False. They get rainbows more frequently.

Weather

Weather

Weather

red, orange, yellow, green, blue, indigo and violet

Weather

Weather

true

Weather

Weather

falling raindrops

Weather

Weather

arc

Weather

Most thunderstorms happen in what season?

Hurricanes start to form when the sun heats the _____.

What time of day do most thunderstorms occur?

Do hurricanes form over warm or cold oceans?

True or False? Lightning can go from one cloud to another.

Name the two seasons that hurricanes usually occur in.

What is the name of the long metal spike attached to the roof of a building and extending to the ground that is used as a safety device to prevent lightning damage?

What is it like inside the eye of a hurricane?

Name the inventor of the lightning rod.

What are the clouds circling the eye of a hurricane called?

Weather Weather

late afternoon or early evening

Weather Weather

late summer or early fall

true

Weather Weather

It is calm and clear.

lightning rod

Weather Weather

Benjamin Franklin

Weather Weather

What is it called when a hurricane moves over the coast onto dry land?

What is the other name that a tornado is sometimes called?

How do meteorologists keep track of hurricanes each year?

Most tornadoes in the U.S. occur in a belt called Tornado _____.

A *willy-willy* is a term for a hurricane or cyclone in what country?

Tornadoes occur most often during what two seasons?

Monsoons happen mainly on what continent?

A waterspout is a tornado that forms over _____.

The winds of a tornado spin in the shape of a _____.

Snow, sleet and hail occur when the air is _____.

twister

Weather

landfall

Weather

Alley

Weather

by naming
the hurricanes

Weather

spring and summer

Weather

Australia

Weather

water

Weather

Asia

Weather

below freezing,
32°F

Weather

funnel

Weather

True or False? Hailstones only fall from cirrus clouds.

A severe snowstorm with strong winds is called a _____.

What do you call ice pellets that form when snow melts and then refreeze when they fall?

True or False? A single snowstorm can drop 40 million tons of snow.

Snow is made of tiny ice _____.

What is the name of the popular motorized sled that people use for recreation and work in snowy regions?

All snowflakes have how many sides?

A sudden fall of snow and ice down a steep slope is called an _____.

What is a ring around the sun or moon created by ice crystals called?

What do we call the icy coating that forms when moisture in the air freezes?

blizzard

Weather

False. Hailstones fall from cumulonimbus clouds.

Weather

true

Weather

sleet

Weather

snowmobile

Weather

crystals

Weather

avalanche

Weather

6

Weather

frost

Weather

halo

Weather

What is the name of permanently frozen ground found mainly in the tundra?

Auroras can be what three colors?

Most precipitation on the tundra is in what form?

What is the coldest continent?

What happens to the ice in the top layer of soil during the summer in the Arctic tundra?

Tropical rain forests grow in warm places near what imaginary line around Earth?

True or False? Polar regions are very windy.

What causes a drought to occur?

What do we call patterns of light in the sky that occur around the North Pole and South Pole?

Dust storms occur where the soil is very _____.

blue, red and yellow

Weather

permafrost

Weather

Antarctica

Weather

snow

Weather

equator

Weather

It melts.

Weather

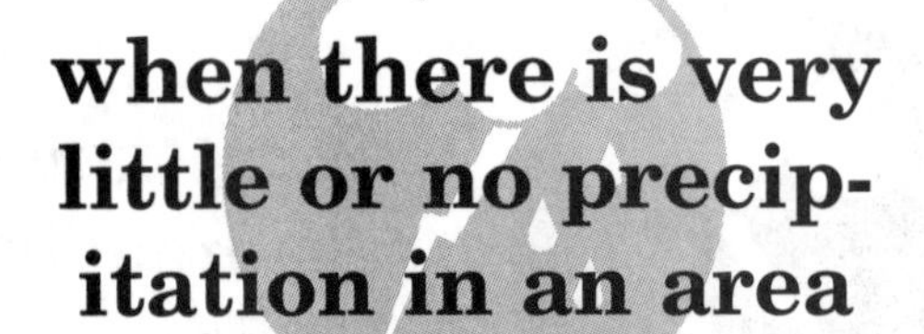

when there is very little or no precip-itation in an area

Weather

true

Weather

dry

Weather

auroras

Weather

An area that receives little rain is called _____.

What direction do changes in the weather usually come from?

What is global warming?

At what temperature does water begin to turn to ice?

The capital *L* symbol on a weather map means "_____ pressure."

What is the difference between a tornado watch and a tornado warning?

What is the high pressure symbol on a weather map?

Where does a stationary front occur?

When meteorologists talk about the combination of wind speed and temperature, they are talking about _____.

What forms when moist air close to the ground is cooled?

west

Weather

desert

Weather

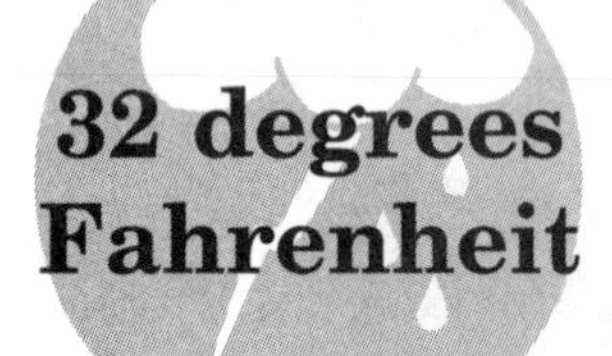

32 degrees Fahrenheit

Weather

Global warming is the gradual heating of Earth.

Weather

A *tornado watch* means "weather conditions are right for a tornado to possibly occur." A *tornado warning* means "a tornado has actually been seen."

Weather

low

Weather

where cold and warm fronts meet

Weather

H

Weather

fog

Weather

windchill

Weather

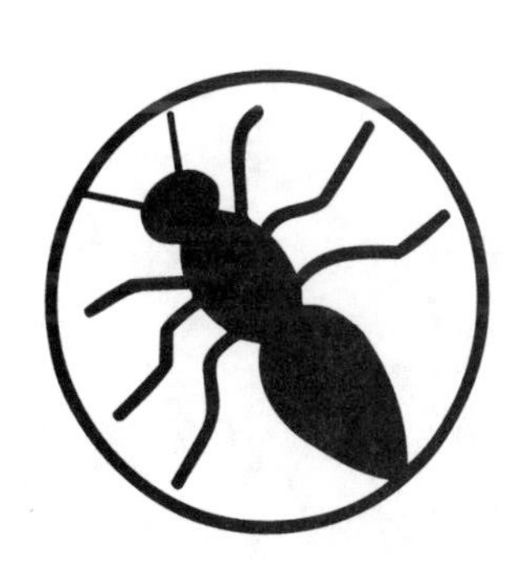

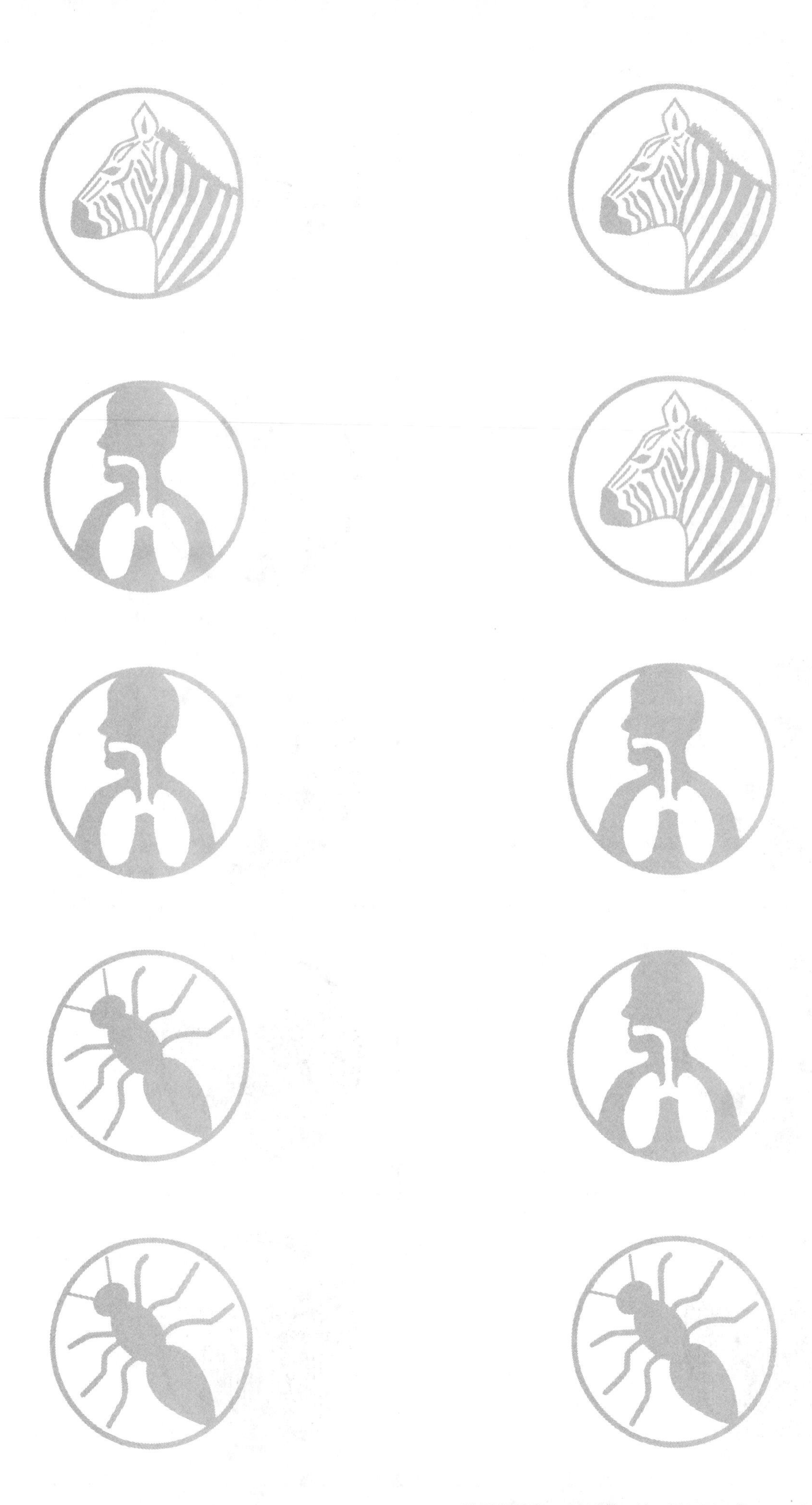